MANAGEMENT OF TUBERCULOSIS

A handbook for clinicians

EDITORS
Alan Street
Emma McBryde
Justin Denholm
Damon Eisen

First published 2012

This work is licensed under the Creative Commons Attribution-NonCommercial 3.0 Unported License. To view a copy of this license, visit http://creativecommons.org/licenses/by-nc/3.0/ or send a letter to Creative Commons, 444 Castro Street, Suite 900, Mountain View, California, 94041, USA.

ISBN 978-1-105-69598-8

Disclaimer

Information in this publication has been obtained from the authors from sources believed to be reliable and up to date at the time of preparation. However, the editors and publisher are not responsible for errors or omissions or for any consequences from application of the information in this book, and make no warranty, express or implied, with respect to the currency, accuracy or completeness of the contents. Application of this information in particular clinical situations remains the responsibility of the medical practitioner.

Publisher and distributor:
Victorian Infectious Diseases Service
Royal Melbourne Hospital, Main Building
Grattan Street, Parkville, Victoria 3050
Australia

Telephone: (03) 9342 7212
Facsimile: (03) 9342 7277
Website: www.vids.org.au

Preface

In the early 2000s, Dr Allen Yung wrote TB management guidelines for use by Victorian Infectious Diseases Service (VIDS) doctors in the clinic and on the ward at The Royal Melbourne Hospital (RMH). They reflected Allen's unparalleled clinical experience in the clinical management of TB, chiefly at Fairfield Hospital and in the latter years of his career at RMH. These guidelines bore the hallmarks of all of Allen's published material: they were carefully thought out and organised, approached problems from the doctor's and patient's perspective, were pitched at junior doctors (but consultants found them very useful as well!), and contained a wealth of information and advice that was often not readily available in published texts. But first and foremost, the guidelines were practical, and they became an indispensable guide to the busy clinician faced with the common and uncommon issues that arise in the course of TB management.

The time has come to update Allen's original guidelines. The aim of providing useful, practical information – the essence of the original document – remains the same, but all chapters have been updated, and new information has been included about many topics. To highlight a few important areas, there is now a substantial section devoted to TB interferon gamma release assays (IGRA), including advantages and limitations of these tests and practical advice about management of patients with discordant IGRA/tuberculin skin test results. The chapter on drug-resistant TB has been completely revised, and includes information about the threat of XDR-TB. A new chapter discusses diagnosis and management of HIV infection and TB, a section has been added about use of BCG for treatment of bladder cancer, and non-tuberculous mycobacterial infections have been consolidated into a single chapter. (We did not include *M. leprae* because comprehensive guidelines exist elsewhere for this specialised infection.)

The target audience for this clinical handbook remains doctors working in RMH and some of the information about public health aspects of management reflects practice in Victoria. We hope, however, that these guidelines could be useful to other doctors working in tuberculosis clinics in Australia. Naturally, some of the material in the handbook will not be applicable to all settings, such as resource-poor countries.

This handbook is not intended to be a comprehensive document and we have not included an extensive list of references as this information can be readily found in specialised TB texts and in international TB guidelines. In the same vein, we have deliberately chosen not to assign an evidence ranking to each of the recommendations. The main focus of the handbook is on clinical management of TB (and other mycobacterial infections); other important areas such as diagnosis, microbiology, public health aspects, TB in poor countries and paediatric TB are not covered in detail. Finally, readers should bear in mind that because of the inevitable delay between the writing and publication of this handbook, it is possible that some of our management recommendations may need to be modified in the light of recently published information.

Our thanks go to all the contributors who gave so willingly of their time and effort, and who have been so patiently waiting for their words to see the light of day!

Alan Street
Emma McBryde
Justin Denholm
Damon Eisen

Acknowledgements

Publication of this handbook would not have been possible without the generous support of the John Burge Trust Fund (see <www.statetrustees.com.au> for further information), a charitable trust which funds activities related to TB community care, treatment, prevention, education and research. This support is gratefully acknowledged, as is the support and encouragement of the staff and patients of the Victorian Infectious Diseases Service TB clinic.

Dedication

These guidelines are dedicated to Dr Allen Yung, the doyen of Melbourne infectious diseases physicians, and the author of the first edition of these guidelines. His legendary teaching has inspired generations of students and doctors, and his particular interest in the clinical management of tuberculosis has had a lasting impact on the care of TB patients in Melbourne and beyond.

Editors
Alan Street, MBBS, FRACP

Emma S McBryde, MBBS, PhD, FRACP

Justin T Denholm, BMed, MPH&TM, MBioethics, FRACP

Damon P Eisen, MBBS, MD, FRACP

Contributing authors
Beverly Biggs, BSc, MBBS, PhD, FRACP, FCRP, FACTH

Ruth Chin, MBBS, PhD, FRACP

Nigel Curtis, MBBS, BA, MA, PhD, Dip TM&H, FRACP

Joseph Doyle, MBBS, FRACP

Sam Hume, MBBS(Hons), FRACP

Chris Lemoh, BMedSci, MBBS, Dip Clin Epi, FRACP

Ben Rogers, MBBS, FRACP

Thomas Schulz, BSc, MBBS, Dip Obs, FRACP

Contents

Preface   *iii*
Acknowledgements   *v*
Dedication   *vi*
List of Abbreviations   *x*

Chapter 1   **Treatment of tuberculosis**   1

Chapter 2   **Management of patients on treatment for tuberculosis**   17

Chapter 3   **Evaluation and treatment of drug-resistant tuberculosis**   37

Chapter 4   **Tuberculosis, pregnancy and perinatal management**   53

Chapter 5   **Diagnosis and management of latent TB infection**   63

Chapter 6   **HIV and tuberculosis**   81

Chapter 7   **Mycobacteria other than tuberculosis**   93

Chapter 8   **BCG vaccination**   115

Index   *130*

List of Abbreviations

AFB	acid-fast bacillus
anti-HBc	hepatitis B core antibody
anti-HBs	hepatitis B surface antibody
ARV	antiretroviral
ALT	alanine aminotransferase
AST	aspartate aminotransferase
ATS	American Thoracic Society
BAL	broncho-alveolar lavage
BCG	Bacille Calmette-Guérin
BTS	British Thoracic Society
BU	Buruli ulcer
CDC	Centers for Disease Control and Prevention, Atlanta GA
COPD	chronic obstructive pulmonary disease
CRP	C-reactive protein
Cs	cycloserine
CT	computerised tomography
CXR	chest X-ray
CYP	cytochrome P450
DoH	Department of Health, Victoria
DOT	directly observed therapy
DOTS	directly observed short course therapy
DST	drug susceptibility testing
E	ethambutol
ESR	erythrocyte sedimentation rate
FBE	full blood examination
GFR	glomerular filtration rate
GI	gastrointestinal
H	isoniazid
HBV	hepatitis B virus
HBsAg	hepatitis B surface antigen
HCV	hepatitis C virus
HRCT	high-resolution computerised tomography
IA	injectable aminoglycoside
IFN	interferon
IGRA	interferon-gamma release assay

INH	isoniazid
IRIS	immune reconstitution inflammatory syndrome
LFT(s)	liver function test(s)
LTBI	latent TB infection
MAC	*Mycobacterium avium* complex
MDR	multidrug-resistant
MIC	minimal inhibitory concentration
NAAT	nucleic acid amplification test
NNRTI	non-nucleoside reverse transcriptase inhibitor
NSAID	non-steroidal anti-inflammatory drug
NTM	non-tuberculous mycobacteria
PAS	para-aminosalicylic acid
PCR	polymerase chain reaction
PI	protease inhibitor
Pro	prothionamide
Q	fluoroquinolones
QFN-GIT	QuantiFERON-TB Gold In-Tube (Cellestis)
R	rifampicin
RBT	rifabutin
RCH	Royal Children's Hospital, Melbourne
RGM	rapidly growing mycobacteria
RMH	Royal Melbourne Hospital
TBM	TB meningitis
TDM	therapeutic drug monitoring
TNF-α	tumour necrosis factor α
TST	tuberculin skin test
U&Es	urea and electrolytes
VIDS	Victorian Infectious Diseases Service, RMH
WHO	World Health Organization
XDR	extensively drug-resistant
Z	pyrazinamide

Chapter 1

Treatment of tuberculosis

1.1	Initial or intensive ('bactericidal') phase of treatment	1
1.2	Continuation ('sterilisation') phase of treatment	4
1.3	Intermittent therapy	6
1.4	Patients with renal impairment	6
1.5	Patients with pre-existing liver disease	7
1.6	Adjunctive use of corticosteroids in tuberculosis	9
1.7	Interrupted or incomplete treatment	13
1.8	Indicators of treatment failure	14

1.1 Initial or intensive ('bactericidal') phase of treatment

Antituberculous treatment is almost always commenced without knowing the antibiotic susceptibilities of the causal organism. The initial regimen usually comprises four drugs (see Table 1.1.1).

1.1.1 HRZE regimen (the standard four-drug regimen)
Drugs: isoniazid (H), rifampicin (R), pyrazinamide (Z) and ethambutol (E)

- Isoniazid is the most effective bactericidal drug.
- Rifampicin and pyrazinamide are the most important sterilising drugs and are thought to act by killing different populations of semi-dormant organisms (persisters).
- Isoniazid and rifampicin are the most effective drugs at preventing the emergence of resistance to other drugs.
- Pyrazinamide has good activity against intracellular organisms and is most active in the first 2 months of treatment; it enables the total duration of treatment to be shortened to 6 months (for fully drug-sensitive infections).
- Ethambutol is a bacteriostatic drug that is given to prevent the emergence of resistance.

Duration of initial (intensive or bactericidal) phase
Minimum duration is 2 months (or 8 weeks). **Pyrazinamide** should be continued until sputum is acid-fast bacillus (AFB) smear-negative or for 2 months, whichever is longer.

Table 1.1.1 Initial dosing of antituberculosis therapy (daily regimen)

Drug	Dose
Isoniazid	10 mg/kg (up to 300 mg) orally, daily
Rifampicin	10 mg/kg (up to 600 mg) orally, daily
Ethambutol	15 mg/kg (up to 1200 mg) orally, daily
Pyrazinamide	25–40 mg/kg (up to 2 g) orally, daily
+ Pyridoxine	25 mg orally, daily

Streptomycin can be administered instead of ethambutol but is now difficult to obtain and is rarely used.

1.1.2 HRZ regimen

Drugs: Three drugs: isoniazid (H), rifampicin (R) and pyrazinamide (Z)

Ethambutol is given to prevent the emergence of additional resistance in the event that isoniazid resistance is already present. Ethambutol is almost always included in the initial regimen because more than 90% of TB in Victoria occurs in individuals born in countries with a high TB incidence and in whom the rate of isoniazid resistance is 5–10% (or higher). Ethambutol is omitted from the initial regimen if the TB isolate is known to be fully susceptible to first-line agents before treatment has been started.

Omission of ethambutol can also be considered if the patient meets the following conditions (indicating a 5% or less likelihood of isoniazid resistance):

- age > 75 and likely to have acquired TB infection before TB treatment became available (early 1950s)
- no previous treatment with antituberculous medications
- birth in Australia, UK, Western Europe or North America, and no prolonged residence in TB-endemic country
- no exposure to a drug-resistant case
- not HIV-infected.

The reason for wanting to omit ethambutol is to avoid the potential for ocular toxicity, especially in elderly patients. However, if there is any doubt, err on the side of caution and use ethambutol.

Ethambutol should not be given to children who are too young to be monitored for visual toxic effects. Ethambutol should also be avoided in patients with impaired renal function. If ethambutol is not used in adults, moxifloxacin should be included in the regimen.

1.1.3 HRE regimen

Drugs: Three drugs: isoniazid (H), rifampicin (R) and ethambutol (E)

Pyrazinamide should be routinely included in the regimen to enable short-course (6 months) therapy to be given, but very occasionally it may have to be omitted (e.g. in a patient unable to tolerate pyrazinamide or in a patient with poorly controlled gout), in which case treatment duration has to be extended to 9 months (see below).

1.1.4 Moxifloxacin

Moxifloxacin is not part of standard regimens, but has excellent *in vitro* activity and is already used to treat patients with multidrug-resistant TB. Preliminary clinical data indicate that it is as effective as ethambutol in the initial phase of standard treatment with a four-drug regimen and that its bactericidal activity is similar to that of rifampicin and isoniazid. For fully drug-sensitive infections, moxifloxacin is indicated in initial regimens for patients unable to take ethambutol. It may also be an effective substitute if either isoniazid or rifampicin cannot be used, but further data are needed before this can be recommended more firmly. Large ongoing randomised controlled trials are investigating its role in abbreviated, 4-month treatment regimens.

1.1.5 Regimens for suspected or proven multidrug-resistant TB

The standard HRZE regimen will provide adequate initial treatment for isoniazid-resistant infections, but not for multidrug-resistant TB (MDR-TB). Fortunately, MDR-TB is still rare (1–2% of notified cases in Victoria and in Australia, although it is likely that this figure will increase in the future), so empirical MDR-TB coverage only has to be considered under special circumstances. For further details regarding indications and regimens for empirical MDR-TB treatment, see Chapter 3, Evaluation and treatment of drug-resistant tuberculosis.

1.1.6 Starting treatment in older patients

Older patients (arbitrarily those aged > 70 years) often experience nausea or vomiting when antituberculous therapy is started. Tolerability can be improved if the drugs are introduced gradually, with a 1–2 day interval between the addition of each agent (given at full dose). Ethambutol is least likely to cause GI intolerance, so it can be started together with isoniazid, followed by rifampicin then pyrazinamide. This strategy does not increase the risk of inducing drug resistance.

1.2 Continuation ('sterilisation') phase of treatment

Quadruple therapy is continued until susceptibility results are known. Ethambutol should be discontinued before the end of the 2-month initial phase if the isolate is fully sensitive to all first-line drugs, providing the patient has responded satisfactorily and has been adherent with therapy. Pyrazinamide should be continued for a minimum of 2 months, and beyond this time if susceptibility results are still pending or the patient remains sputum smear positive. In the continuation ('sterilisation') phase the regimen can be simplified.

1.2.1 *M. tuberculosis* fully sensitive to all first-line drugs

If the isolate is sensitive to all four drugs, **pyrazinamide** and **ethambutol** are ceased, and double therapy with rifampicin and isoniazid is continued to complete a 6-month course of treatment (unless there is an indication to treat for longer than 6 months – see below).

This regimen in shorthand is **2 HRZE/ 4 HR** (the regimen **2 HRZ/ 4 HR** is also acceptable when ethambutol is not used in the initial phase, as discussed above).

This standard regimen (2 HRZE/ 4 HR or 2 HRZ/ 4 HR) should be used in all patients with fully sensitive *M. tuberculosis*. It has the ability:

1. to cure patients rapidly
2. to cure the great majority of patients with bacilli initially resistant to isoniazid
3. to prevent therapeutic failure due to the emergence of acquired resistance.

For *M. tuberculosis* resistant to one or more first-line drugs refer to Chapter 3, Evaluation and treatment of drug-resistant tuberculosis.

Caveats to the standard 2 HRZE/ 4 HR (or 2 HRZ/ 4 HR) regimen
- If pyrazinamide is not used in the initial phase, the total duration should be at least 9 months; the regimen in shorthand is 2 HRE/ 7 HR. (The term 'short course' therapy implies a duration of 6 months; it can be used only if pyrazinamide is included in the initial 2 months of treatment.)
- Isoniazid should be used for the full duration of any treatment regimen unless there are specific reasons against it, such as adverse effects or bacterial resistance.
- Rifampicin is an essential drug for any regimen less than 12 months.
- In the standard regimen the 6-month course should be regarded as a minimum period. A 7-month continuation phase should be given in patients with cavitary disease and who have positive sputum 2 months into initial

treatment. Patients not treated with pyrazinamide in the initial regimen should also have a 7-month continuation phase.

1.2.2 Duration of treatment of extra-pulmonary TB due to fully sensitive *M. tuberculosis*

Most forms of drug-sensitive extrapulmonary TB can be treated with a standard 6-month short course regimen, unless the disease is extensive or complicated. Specific forms of extrapulmonary TB that require a longer course of treatment are:

- disseminated (miliary) tuberculosis – 9 to 12 months
- tuberculous meningitis – 12 months
- skeletal tuberculosis – depending on the extent of disease, 6 to 9 months.

1.2.3 If no *M. tuberculosis* is cultured

Sometimes the diagnosis of TB is not confirmed by a positive culture, for example if:

- specimens have not been sent for TB culture (e.g. excised lymph node tissue put in formalin)
- few organisms are present
- the tuberculous disease is inactive
- the pre-treatment diagnosis of TB is wrong and the patient has another condition altogether
- the patient has an atypical mycobacterial infection caused by a difficult-to-culture organism.

If AFBs are visible in smears or tissue sections but are not cultured, one or more of the following molecular tests is recommended:

- TB PCR
- generic mycobacterial PCR – if non-tuberculous infection suspected
- RNA polymerase B TB PCR (for rifampicin resistance mutations) – if MDR-TB suspected.

In patients with culture-negative pulmonary disease, the major indicators of response to empiric TB treatment are radiological and clinical.

If there is symptomatic and/or CXR improvement after 2 months of treatment with HRZE, this suggests that the disease is active, and treatment should be continued. The American Thoracic Society (ATS) guidelines recommend HR for only another 2 months (that is, a 4-month treatment course), but in VIDS we take a more conservative approach, treating for longer and also covering the possibility of isoniazid-resistant disease (but continuing isoniazid throughout). The regimen used in VIDS is:

2 HRZE/ 7 HRE

Failure of the CXR to improve is strongly suggestive that the abnormality is the result of either previous (not current) tuberculosis or another process. TB therapy can usually be stopped since the patient will have been adequately treated for latent TB at this stage.

1.3 Intermittent therapy

WHO and other international authorities recommend routine use of intermittently administered, directly observed short course therapy (DOTS) for all tuberculosis cases. Practice in Australia varies from state to state: in Victoria, patients' treatment is closely supervised by the TB Control Program and the treating clinician, and DOTS is provided on a case-by-case basis for patients with anticipated or documented adherence problems.

Duration of therapy and treatment outcome are the same for DOTS as for daily therapy. In shorthand the standard DOTS regimen is 2 HRZE/ 4 H_3R_3. Twice-weekly intermittent therapy is recommended in the United States but not in the WHO or UK National Institute of Health and Clinical Excellence (NICE) guidelines. Daily therapy is usually given initially, followed by a switch to intermittent therapy after 2 to 8 weeks.

The practice of giving one or two drugs daily and others intermittently (three times per week) is inadvisable and dangerous if the daily dose is not accompanied by another drug also given daily, because of the risk of developing resistance to a single daily drug.

1.4 Patients with renal impairment

Isoniazid, rifampicin, pyrazinamide, prothionamide and ethionamide are eliminated almost entirely by hepatic metabolism or by biliary excretion.

Isoniazid and rifampicin

Isoniazid and rifampicin are the safest antituberculous drugs in patients with renal impairment. Isoniazid can be given in standard doses in renal impairment, including dialysis patients. Some sources recommend dose reduction for rifampicin but the ATS 2003 guidelines support full-dose rifampicin and this is the practice in VIDS.

Pyrazinamide
There is disagreement regarding pyrazinamide dosing in patients with moderate to severe renal impairment. British Thoracic Society (BTS) guidelines suggest that it can be given in standard dosage whereas the ATS guidelines suggest dosage reduction to 25–35 mg/kg three times a week if the GFR is < 30 mL/min. We suggest following the specific dosing recommendations in Therapeutic Guidelines: Antibiotic, which advise pyrazinamide dose reduction only with severe renal impairment (GFR < 10 mL/min) (see Table 1.4.1).

Streptomycin, amikacin and other aminoglycosides
Aminoglycosides are excreted exclusively by the kidneys. It is best to avoid these drugs altogether in patients with renal impairment, but if their use is absolutely essential, monitoring of levels is done to guide appropriate dosing.

Ethambutol
Ethambutol is excreted predominantly by the kidney. Although dosage can be adjusted according to degree of renal impairment (see Table 1.4.1), the availability of moxifloxacin means that ethambutol should rarely need to be used in the setting of renal impairment.

Ethionamide/prothionamide
These drugs are metabolised and not renally excreted.

- For GFR > 10 mL/min — no change in dose
- For GFR < 10 mL/min — 5 mg/kg/day

1.5 Patients with pre-existing liver disease

- Patients with pre-existing liver disease or abnormalities of liver function (e.g. due to chronic hepatitis B and C, alcohol or other forms of liver disease) are at increased risk of additional liver damage from antituberculous drugs.
- Regular monitoring of LFTs and close clinical follow-up is required for these patients – weekly for 2 weeks then fortnightly for 2 months with subsequent monthly follow-up if stable.
- Rifampicin, isoniazid and pyrazinamide are all potentially hepatotoxic.
- Isoniazid and rifampicin can generally be used with careful monitoring (as discussed in more detail below), although the combination is best avoided in patients with severe liver disease.
- Pyrazinamide has the highest rates of hepatotoxicity – it can also be used with careful monitoring but is best avoided in patients with severe liver disease.

Table 1.4.1 Dosing of antituberculous drugs in patients with renal impairment.

Drug	Adjustment for renal failure by GFR (mL/min)			Doses for dialysis		
	> 50	10–50	< 10	HAEMO	CAPD	CRRT
Isoniazid	normal	normal	normal	normal, dose after dialysis	normal	normal
Rifampicin	normal	normal	50–100% at normal interval	as for GFR < 10	as for GFR < 10	as for GFR < 10
Pyrazinamide	normal	normal	50–100% at normal interval	25–30 mg/kg after each dialysis	normal	normal
Ethambutol (see Notes)	normal	66–100% daily	100% 48-hourly	15 mg/kg after each dialysis	as for GFR < 10	15 mg/kg 24-36-hourly
Ciprofloxacin	normal	50–75% 12-hourly	100% daily or 50% 12-hourly	as for GFR < 10, dose after dialysis	as for GFR < 10	200–400 mg 12-hourly
Moxifloxacin	normal	normal	normal	normal, dose after dialysis	normal	normal
Amikacin (see Notes)	50–100% 24-hourly	50% 24-hourly to 30% 48-hourly	30% 48-hourly	seek expert advice		
Streptomycin (see Notes)	normal	50–100% daily	100% 72–96-hourly	50% dose after each dialysis	100% 72–96 hourly	dose for GFR 10–50
Cycloserine	normal	100% 24-hourly	100% 36–48-hourly	as for GFR < 10	as for GFR <10	as for GFR 10–50

Adapted from: Antibiotic Expert Group. Therapeutic guidelines: antibiotic. Version 14. Melbourne: Therapeutic Guidelines Limited; 2010.

Notes
- HAEMO = haemodialysis; CAPD = continuous ambulatory peritoneal dialysis; CRRT = continuous renal replacement therapy.
- Ethambutol, amikacin and streptomycin should be avoided in patients with renal impairment unless there are no alternatives.
- Monitoring of amikacin and streptomycin levels is recommended to determine precise dosage requirement.

- Drugs that can be used without additional monitoring in liver disease are aminoglycosides, capreomycin, cycloserine, ethambutol and fluoroquinolones.

1.5.1 Patients with pre-existing impaired liver function
- If the LFTs are mildly abnormal and likely to be due to TB, disregard but monitor carefully.
- If the abnormal LFTs are due to chronic liver disease, use either rifampicin or isoniazid at a lower dose initially (e.g. 450 mg and 200 mg respectively for adults with body weight ≥ 50 kg) with pyrazinamide and ethambutol. The other drug, i.e. either isoniazid or rifampicin, may be added after 1 or 2 weeks if there is no deterioration of liver function and no symptoms of hepatitis.
- Increases in aminotransferases are common. Drugs should be stopped if aminotransferases are > 5× the upper limit of normal, or ≥ 3× in the presence of symptoms. Otherwise, treatment can be continued, providing monitoring is more frequent and information about symptoms of hepatitis is reinforced.
- More information about management of deterioration in liver function while on therapy is included in Chapter 2, Management of patients on treatment for tuberculosis.

1.5.2 Patients with pre-existing severe liver disease
- The World Health Organization (WHO) suggests either 2 SHRE/ 6 HR or 9 RE, or 2 SHE/ 10 HE (S = streptomycin) for patients with liver failure.
- ATS guidelines suggest HRE for 9 months (with cessation of ethambutol if fully sensitive), and in patients with cirrhosis, RE with a fluoroquinolone or cycloserine for 12–18 months.

Both these regimens avoid pyrazinamide, but still entail administration of isoniazid and/or rifampicin.

If hepatotoxic medications must be avoided altogether (e.g. hepatic decompensation or severe LFT abnormalities), a suggested regimen is ethambutol plus moxifloxacin plus an injectable aminoglycoside plus (possibly) cycloserine.

1.6 Adjunctive use of corticosteroids in tuberculosis

Although corticosteroids are a well recognised risk factor for TB reactivation in patients with latent TB infection, they are also of proven benefit in some forms of TB when given as an adjunct to antituberculous therapy.

1.6.1 Considerations before use

Exacerbation of underlying chronic hepatitis B and strongyloidiasis

These conditions share a common geographical distribution with TB. Unmonitored corticosteroid use can lead to a flare of hepatitis B or to the development of disseminated strongyloidiasis. Hepatitis B surface antigen status should be known in all patients before use. Strongyloides serology should be requested in all refugees (if not done previously) and in overseas-born patients with unexplained gastrointestinal, skin or respiratory symptoms, or with unexplained eosinophilia.

Side-effects of corticosteroids

Diabetes mellitus, osteoporosis and gastrointestinal ulceration are some of the many side-effects of corticosteroids. Blood glucose levels should be checked periodically, especially during the initial, higher-dose phase of treatment, and more intensively in those with pre-existing diabetes. The risk of osteoporosis is low with short-term corticosteroid therapy but vitamin D levels should be checked and corrected if low, and calcium intake should be optimised. Patients with a past history of peptic ulcer disease or those on regular NSAID treatment should be considered for gastro-protective therapy with a proton pump inhibitor.

1.6.2 Definite indications

Corticosteroids should be used as adjunctive therapy to antituberculous treatment in the following situations.

Pericarditis

Adjunctive corticosteroid therapy for tuberculous pericarditis is recommended. Multiple smaller trials have demonstrated benefits including enhanced resolution of effusion and lower mortality, although not with clear statistical significance. Meta-analysis of four of the trials showed benefits including reduced mortality and reduced requirement for repeat pericardiocentesis, again with very broad confidence intervals. The authors of the meta-analysis concluded that corticosteroids have an important clinical benefit although trials to date have been too small to demonstrate this effect beyond doubt. Corticosteroids have not been shown to prevent the later complication of constrictive pericarditis. Input from the cardiology and cardiothoracic surgical service should be requested in every patient with suspected tuberculous pericarditis.

Recommended dosing: prednisolone 60 mg daily for 4 weeks, 30 mg daily for 4 weeks, 15 mg daily for 2 weeks, 5 mg daily for 2 weeks (total 12 week course)

Meningitis (TBM) and intracerebral tuberculoma
Adjunctive corticosteroid therapy is of proven benefit in TBM. A 2008 meta-analysis of seven trials including 1140 participants demonstrated a significantly reduced risk of death and disabling residual neurological deficit. This benefit was applied to all stages of TBM (Medical Research Council stage I–III). CNS damage in TBM results from occlusion of ventricular foramina (causing hydrocephalus), vasculitis of perforating blood vessels (causing cerebral infarction) and involvement of cranial nerves. Corticosteroids exert their beneficial effect by reducing the inflammation in the sub-arachnoid space that is responsible for these complications.

In TBM patients administered corticosteroids, symptoms such as fever, headache, malaise and delirium may improve more rapidly, and CSF may normalise (in protein level and cell count) more quickly than otherwise. All these clinical and laboratory parameters may rebound when corticosteroids are discontinued, even when the dose has been carefully tapered. Despite an oft-cited concern that steroid-induced suppression of meningeal inflammation could reduce CNS penetration of antituberculous drugs, there is no difference in CSF levels of commonly-used agents with or without concomitant corticosteroids.

Most experts would also advise routine use of adjunctive corticosteroids in patients with intracerebral tuberculoma, even though corticosteroids are not of proven benefit in this situation.

Recommended dosing: dexamethasone 12–16 mg daily for 3 weeks then tapered over the following 3 weeks **or** prednisolone 60 mg daily for 3 weeks then tapered over the following 3 weeks (From Prasad & Singh, 2008)

HIV-associated immune reconstitution syndrome
HIV-infected patients who are being treated for TB can develop an exaggerated ('paradoxical') inflammatory reaction when antiretroviral therapy is commenced. Clinical complications such as worsening respiratory distress or increased intracranial pressure can develop as a result (see Chapter 6, HIV and tuberculosis). Risk factors are a short interval (< 4 weeks) between starting TB and HIV therapy, a low CD4 cell count (< 100 per microlitre) and disseminated disease. In a randomised controlled trial, corticosteroid therapy was shown to be effective treatment for this condition.

1.6.3 Possible indications
There is limited evidence for the benefit of corticosteroids as adjunctive therapy to antituberculous treatment in settings other than TBM and TB pericarditis. In these other settings, adjunctive corticosteroids are sometimes used in

exceptional circumstances, such as in the seriously ill patient where more rapid resolution of symptoms may be beneficial. A clinical decision about corticosteroid use in these settings should be made on a case-by-case basis.

Endobronchial tuberculosis
In paediatric patients, enlarged intrathoracic lymph nodes can compress bronchi leading to lobar or segmental collapse-consolidation and respiratory distress; this often improves with use of corticosteroids. Bronchial obstruction is also a complication of endobronchial TB, but there is no evidence of benefit for corticosteroid therapy in this situation.

Tuberculous pleural effusion
A 2007 meta-analysis reported that adjunctive use of corticosteroids in TB pleural effusion was associated with reduction in size of the effusion at 4 weeks, earlier resolution of symptoms and less residual pleural thickening, but the magnitude of the benefit was small. Studies have not consistently demonstrated long-term reduction in morbidity or preservation of respiratory function after completion of antituberculous treatment.

Extensive pulmonary disease
There have been numerous controlled studies of the possible role of corticosteroids in the treatment of pulmonary tuberculosis. Some have demonstrated benefits early in the course of disease treatment, including a more rapid clinical and radiological improvement, but this benefit is brief, with no demonstrated difference between steroid and control groups at 3–6 months. Studies have also described rebound worsening of clinical status with corticosteroid withdrawal. On the basis of this evidence, corticosteroids cannot be recommended for the sole indication of pulmonary disease. However, some experienced clinicians would use corticosteroids for the occasional patient who presents with extensive pulmonary disease, respiratory distress and severe systemic symptoms such as marked weight loss ('galloping consumption').

Ureteric tuberculosis
Renal tract tuberculosis involving the ureter can lead to obstruction, hydronephrosis and renal insufficiency. Primary management of mechanical renal obstruction should be via early urological intervention with ureteric stenting or percutaneous nephrostomy. Older trials have demonstrated that corticosteroids can reduce ureteric stenosis and stricture in patients with ureteric tuberculosis, but their concurrent use with modern urological interventions has not been investigated.

Other indications
Corticosteroids have been used in a variety of other forms of tuberculosis. Examples include gastrointestinal TB (to prevent stricture formation), vertebral TB with involvement of the epidural space (to reduce the risk of spinal cord or nerve root compression) and tuberculous arthritis (to prevent joint ankylosis). Despite a weak or non-existent supportive evidence base, there are some compelling case reports in the literature and anecdotes from experienced clinicians that are hard to ignore, so corticosteroid use in these situations should not be dismissed entirely.

1.7 Interrupted or incomplete treatment

Accidental or necessary interruptions to TB treatment are common. After a patient has an interruption to TB treatment a decision must be made about the best course of further treatment. In some circumstances it may be possible to restart the interrupted regimen, whereas a completely new course of treatment is required in other instances.

A TB regimen can be considered as a requisite number of days (or doses) of therapy. If a treatment interruption of any length occurs, the usual course of action is to prolong the regimen to compensate for any missed doses.

There is no evidence basis for recommendations relating to treatment interruption, and the definition of a treatment interruption requiring intervention varies widely between guidelines. Factors related to the patient's immune status and extent of disease must also be considered. The recommendations below are only for isolates treated with first-line agents. In some situations, these recommendations advocate a more cautious approach than that found in international guidelines, as described below.

1.7.1 Durations of interruptions of therapy requiring further intervention
- Interruption for ≥ 14 days during initial phase of therapy
- Interruption for ≥ 1 month during the continuation phase of therapy

1.7.2 Restarting a new treatment course
The TB treatment regimen should be *restarted* from the beginning, without compensation for the duration of previous therapy taken, in the following situations:

- Interruption of > 14 days in initial phase of therapy, or

- Interruption of > 2 months in continuation phase. (Note: This differs from WHO guidelines, which recommend only restarting in this situation if the patient was also smear positive before or after interruption.)

1.7.3 Recommencing the previous regimen

The TB treatment regimen can be recommenced in circumstances other than those listed above, the aim being to provide the patient with the same total number of doses of treatment that were originally planned.

For patients who interrupt therapy for less than 2 months in the continuation phase, the WHO recommendation would be for the patient to receive treatment for a total of 6 months (180 daily doses, or equivalent) by adding the lapsed period to the originally planned stopping date. However, our practice in this situation (not evidence-based) is often more conservative, and we may extend the continuation phase to provide a total duration on antituberculous therapy of 9 months. We may also consider restarting rather than recommencing treatment in patients who were smear or culture positive at the time of the treatment discontinuation, in immunosuppressed patients, and in patients at higher risk of treatment failure (e.g. with extensive or cavitary disease, or severe underlying lung pathology such as silicosis), depending on factors such as whether treatment was ceased early in the continuation phase.

Key steps in restarting or recommencing treatment
- Identify and address the issues that contributed to the interruption of treatment.
- Obtain new sputum samples to determine smear status and to undertake repeat sensitivity testing.
- Discuss arrangements for enhanced supervision or directly observed therapy (DOT) with the public health TB program and implement other measures to enhance adherence (e.g. support for financial, accommodation, transport, or substance abuse problems).

1.8 Indicators of treatment failure

1.8.1 Definition of treatment failure

Treatment failure, an entity distinct from disease relapse, is defined by failure of resolution or recrudescence of active disease while on therapy, rather than recurrence after completion of treatment.

- Positive TB cultures while on treatment any time after completion of the 4th (ATS) or 5th (WHO) month of antituberculous therapy

or

- Reversion from a negative culture to a positive culture while on treatment any time after completion of the 3rd month of antituberculous therapy.

In addition, treatment failure should be considered if cultures are pending and the individual has clinical deterioration or radiological deterioration suggestive of treatment failure. After 3 months of therapy containing HR, 90–95% of patients should be culture negative. Positive cultures after this time should lead to close monitoring for impending treatment failure.

1.8.2 Common causes of treatment failure
- Drug resistant isolate
- Poor medication compliance – the most common reason
- Low drug levels. Consider:
  - drug interactions
  - poor absorption
  - incorrect dose calculation
  - dispensing or administration errors.

1.8.3 Management of treatment failure
- Obtain repeat samples for culture and sensitivity testing. The laboratory should be requested to check susceptibility of the most recent isolate to second-line, as well as first-line, agents.
  - Pulmonary TB: three consecutive daily sputum samples (spontaneous or induced); if unobtainable, organise bronchoscopy.
  - Extrapulmonary TB: renewed attempts should be made to obtain appropriate specimens for AFB smear, culture and susceptibility testing.
- If the patient is clinically stable, the current regimen may be continued until new susceptibility results are available to guide the choice of medications.
- If the patient is clinically deteriorating the regimen should be modified as per the recommendations in Chapter 3, Evaluation and treatment of drug-resistant tuberculosis.

- Directly observed therapy (DOT) should be instituted if the individual is not currently under this method of care – discuss with the public health TB program.
  - Address social and other issues that may have contributed to treatment failure.
  - For drug-sensitive infections, extend duration of therapy to 6 months after culture conversion.

References and further reading

Antibiotic Expert Group. Mycobacterial infections. In Therapeutic Guidelines: antibiotic. Version 14. Therapeutic Guidelines Limited, Melbourne, 2010, pp. 171–88.

Centers for Disease Control and Prevention. Treatment of tuberculosis. American Thoracic Society, CDC and Infectious Diseases Society of America. MMWR Recommendations and Reports. 2003; Vol 52 No. RR-11.

Joint Tuberculosis Committee of the British Thoracic Society. Chemotherapy and management of tuberculosis in the United Kingdom: recommendations 1998. Thorax 1998; 53:536–48.

National Institute for Health and Clinical Excellence. Tuberculosis: Clinical diagnosis and management of tuberculosis, and measures for its prevention and control. Available at: <www.nice.org.uk/nicemedia/pdf/CG033FullGuideline.pdf>.

Prasad K, Singh MB. Corticosteroids for managing tuberculous meningitis. Cochrane Database of Systematic Reviews 2008, Issue 1. Art. No.: CD002244. DOI: 10.1002/14651858.CD002244.pub3.

WHO. Treatment of tuberculosis guidelines. 4th edition. World Health Organization, Geneva, 2010.

Chapter 2

Management of patients on treatment for tuberculosis

2.1	Site of care and isolation	17
2.2	Notification	18
2.3	Contact tracing	18
2.4	Before starting therapy	19
2.5	Routine inpatient monitoring	21
2.6	Discharge from hospital	21
2.7	On-treatment review in the outpatient clinic	22
2.8	Post-treatment follow-up	24
2.9	When patients develop problems	25
2.10	Usual response to treatment	33

2.1 Site of care and isolation

Commencement of therapy as an outpatient can be considered for patients who are non-infectious (e.g. lymph node TB or truly asymptomatic pulmonary TB) who are otherwise clinically stable and who have satisfactory social circumstances.

Patients with bacteriologically confirmed or clinically suspected active pulmonary TB should be admitted to hospital and placed in a single TB isolation room maintained under negative pressure.

Patients moving outside the isolation room should wear N95 masks. Where possible investigative procedures for these patients should be scheduled at times when they can be performed rapidly and when patients are not held in crowded waiting areas for long periods.

The number of healthcare workers (HCW) or visitors entering an isolation room should be kept to a minimum. All persons entering an isolation room should wear an N95 mask.

TB patients should be educated about the mechanisms of TB transmission, and the need to cover their mouths and noses when coughing or sneezing to minimise the droplet spread of mycobacteria in expelled air.

Most patients are no longer infectious 2 weeks after commencing an HRZ-containing regimen as cough has usually lessened markedly and most of the bacilli are dead. However one should be more guarded with a patient who continues to have a hacking cough, has cavitary disease or has plentiful AFB on sputum smears. This applies particularly to the multidrug-resistant case. In practice, we tend to err on the side of caution and maintain isolation in hospital until the patient is discharged, unless the admission is prolonged.

2.2 Notification

All patients started on full antituberculous treatment, whether or not the diagnosis of TB has been confirmed at the time, must be notified to the relevant state or territory public health TB program. In Victoria this is the TB Control Program, Department of Health (DoH). Tuberculosis is a group B disease, requiring written notification within 5 days.

- For some situations involving patients with smear positive pulmonary TB, contact tracing may be a matter of urgency, in which case notify the TB program by telephone (in Victoria, on 9096 5144 or 1300 651 160). Examples include:
  - close contact with infants and young children
  - recent travel on an international flight
  - exposure in congregate settings, such as prison, nursing homes and other care facilities.
- In Victoria, complete the Enhanced Notification of TB form – this is available online at <http://ideas.health.vic.gov.au/notifying/what-to-notify.asp>.
- Copy notification form and file in the patient's history.
- Return the notification form – in Victoria to DoH by either:
  - Mail (pre-printed envelopes available) to
    Communicable Diseases Control (Public Health Branch)
    Victorian Government Department of Human Services
    Reply Paid 65937
    Melbourne VIC 8060
  - Fax to 1300 651 170.

2.3 Contact tracing

Tracing of the index patient's community contacts is the responsibility of the relevant state or territory public health TB program, and should not be initiated by the hospital medical staff unless there is an untreated symptomatic contact who may need more urgent management. Concerned family, social or work contacts who request 'testing' should be referred to the TB program public health nurse who has been assigned to the case.

In contrast to contacts in the community, if hospital staff or patients have occupational exposure to a hospital patient with unrecognised TB before institution of airborne precautions, contact tracing is the responsibility of the hospital. It is usually carried out by infection control personnel, in consultation with relevant specialists such as infectious diseases or respiratory physicians or microbiologists. The relevant TB program should be notified of these measures.

Patients receiving treatment of latent TB do not need to be notified.

2.4 Before starting therapy
2.4.1 Educate patient and family
- For non-English-speaking patients, use an interpreter *in person* – a telephone interpreter is usually unsatisfactory.
- Give information about TB – excellent prognosis with treatment, likelihood of transmission (reassure if non-pulmonary TB), requirement for notification to public health authorities.
- Provide detailed information about treatment (for inpatients, involve the ward pharmacist to reinforce explanations):
  - importance of adherence
  - specific dosing instructions (use aids such as tablet cards and sheets) – number of tablets, frequency, timing in relation to food
  - side-effects – specifically hepatitis (describe symptoms, and instruct patient to stop all medications if they develop), eye toxicity (instruct patient to stop medications if altered colour vision or blurred vision develops), early GI upset, orange urine, arthralgias, allergy
  - written information (for example DoH pamphlets in Victoria) about TB and TB drugs in appropriate language (if available).
- Provide clinic/hospital contact details.

2.4.2 Check for drug interactions
Rifampicin
- Oral contraceptive pill – advise patient to use an alternative contraceptive method.
- Prednisolone – if being given for a pre-existing condition, a rule of thumb is to double the prednisolone dose (discuss with treating doctor or parent unit beforehand).

Table 2.5.1 Inpatient monitoring schedule.

	Baseline	Weekly	On discharge
LFTs	✓	✓	✓
FBE	✓		✓
U&Es	✓	If on amikacin	✓
ESR	✓		
HIV	✓		
HCV	✓		
HBV	✓		
Vitamin D	✓		

- Warfarin – increase frequency of INR monitoring until the warfarin dose is stabilised.
- There are many other potentially important interactions – for details see texts such as UpToDate or consult ATS guidelines, and if in doubt, contact Pharmacy.

Isoniazid
- Phenytoin – monitor levels.

2.4.3 Baseline laboratory tests
- Full blood examination (FBE), urea and electrolytes (U&Es), and LFTs
- HIV testing – after appropriate provision of information
- Hepatitis B serology (HBsAg, anti-HBs, anti-HBc) and hepatitis C antibody for all overseas-born patients
- 25-OH vitamin D level (and provide replacement if low)
- Other immigrant and refugee health testing such as *Strongyloides* and schistosomiasis serology if not done previously, as appropriate.

2.4.4 Baseline visual acuity and colour vision
- Before starting ethambutol, check and document baseline visual acuity and colour vision.
- Refer to ophthalmology if there are pre-existing visual problems or a history of significant eye disease – otherwise, advice from our ophthalmologists is that formal monitoring by an ophthalmologist is *not* required, either at baseline or on treatment.

2.5 Routine inpatient monitoring

A second **CXR** is not routinely necessary until 3 months after the initial one.

After 2 weeks of chemotherapy, **sputum smear** should be performed once a week if it continues to be positive.

Serum **uric acid** should be performed if arthralgia occurs while on pyrazinamide.

2.6 Discharge from hospital

The decision about when a patient with smear-positive pulmonary TB can be discharged from hospital is not always straightforward, and should be made on a case-by-case basis by the responsible consultant physician, with input from other specialists and public health authorities if necessary. There is no 'magic formula' that can be applied; in practice, patients usually meet the following criteria around 2 weeks after starting TB treatment, but some patients can be safely discharged earlier than this while others will need a longer admission.

2.6.1 Ideal criteria to be met before patient is discharged

- There is no uncontrolled associated illness or condition that is likely to impede progress.
- Cough is controlled.
- Sputum quantity is reduced and the patient understands the possibility of infecting contacts.
- In patients with smear-positive pulmonary TB, there should be a reduction in the number (but *not* necessarily absence) of organisms seen on smears before discharge.
- In patients with smear-positive pulmonary TB who will have contact in the home with children < 5 years of age or immunocompromised persons, or who are to be transferred to another short- or long-term care facility, smears should ideally be negative before discharge but factors such as duration of therapy (≥ 2 weeks) and resolution of cough may moderate this recommendation.
- The patient has satisfactorily tolerated a drug regimen that can be managed on an outpatient basis.
- The patient understands the disease and the need for compliance.
- Home conditions are satisfactory, and reliable home supervision is available, ideally with at least one family member or other appropriate person who appreciates the disease and the need for patient compliance.

- Follow-up arrangements have been made for routine clinical reviews and supply of drugs, and for visits by an appropriate nurse or health worker as necessary.

2.6.2 On discharge, the following steps should be taken
- Inform the public health TB program (in Victoria, the TB Control Program on 1300 651 160 or fax 1300 651 170).
- Ring the referring doctor or local medical officer and inform of diagnosis, treatment, present state of health and plan of management.
- Complete the discharge summary and ensure a copy is faxed to the referring doctor and to the public health TB program.
- Make a follow-up appointment at the outpatient clinic (usually 2 to 4 weeks after discharge), with an interpreter if needed.
- Ensure that the patient has a sufficient supply of antituberculous drugs to last until the first follow-up appointment. This may require a specific note on the discharge summary prescription if the default hospital pharmacy practice is to dispense a fixed supply of medications (e.g. only for 5 days) at discharge.
- Ensure full cooperation with the responsible person at home.
- In patients with complicated or special management issues, ensure that the management plan is clearly outlined on the discharge information sheet, or preferably discuss the case with a TB Clinic consultant.

2.7 On-treatment review in the outpatient clinic
2.7.1 Frequency of review
If treatment has been commenced in the outpatient setting, patients should be reviewed 2 weeks after initiation, then monthly thereafter.

For patients who do not attend scheduled appointments, inform the TB program public health nurse, who will coordinate a response with the clinic staff – the patient should be contacted and given another appointment date and, if medications are running low, a prescription should be given via the public health nurse.

2.7.2 Clinical monitoring
Related to disease
- Recurrence of cough
- Recurrence of haemoptysis
- Weight loss.

Related to drugs
- Fever and/or rash
- Gastrointestinal symptoms
- Joint pains
- Jaundice
- Visual symptoms – while on ethambutol ask patient about visual symptoms, and check visual acuity and colour vision at each visit.

2.7.3 Drug compliance
Check if the patient has been compliant with the medications if directly observed therapy is not administered. The quantity remaining in the bottles is an unreliable indicator. The urine colour should be red in those taking rifampicin. A urine test for isoniazid can also be done. If there is any doubt with compliance, discuss this with the TB program public health nurse, who will arrange for closer supervision in the community. In selected cases, directly observed therapy may be indicated.

2.7.4 Blood tests
Some TB clinicians choose to monitor LFTs in all patients on TB therapy, but this is not our practice at VIDS (in line with recommendations in international guidelines) and we restrict regular LFT monitoring to the following patients:
- upper gastrointestinal symptoms (anorexia, nausea, vomiting, abdominal pain)
- older than 60 years
- abnormal baseline LFTs
- excess alcohol intake
- concomitant hepatotoxic medication
- chronic HBV infection
- chronic HCV infection (5× increased risk of hepatitis)
- HIV infection (4× increased risk – 14× if HCV coinfected)
- pregnancy, and 3 months post-partum.

FBE and U&Es should be performed only if clinically indicated.

2.7.5 Microbiology
Pulmonary TB
- Check sputum specimens monthly until smear and culture negativity is documented.

- If patients are unable to produce sputum, obtaining respiratory secretions (e.g. induced sputum) entails some inconvenience and discomfort and is only indicated if the patient has MDR-TB, or is otherwise not responding satisfactorily on clinical or radiological grounds. If the patient has a drug-sensitive infection and is doing well, this is not necessary.

Extrapulmonary TB
- If there is a delayed treatment response, obtain additional specimens if possible.

2.7.6 Radiology
For patients with pulmonary TB, CXR should be performed at 3 months, and on completion of therapy.

For patients with extrapulmonary TB, the need for radiological follow-up will depend on the site involved. For instance, routine assessment of radiologic progress is indicated for CNS tuberculoma, but not for lymph node disease.

2.7.7 Communication
Write to the local doctor and the public health TB program at least every 3 months. In addition, if the patient has special problems the TB program nurse should be contacted (in Victoria, on 1300 651 160). In Victoria, a Treatment Outcome Form will be sent by the TB Control Program to the treating physician for completion at the close of treatment. Other units involved in care should also be updated on progress.

2.8 Post-treatment follow-up
2.8.1 Pulmonary tuberculosis
Although routine follow-up of patients after an uncomplicated course of treatment for fully drug-sensitive pulmonary TB is no longer recommended in US and UK treatment guidelines, our practice in VIDS is to review patients at 6 months, 12 months and 24 months, with a repeat CXR at each visit. A letter should be sent to the public health TB program when the patient is discharged from the clinic.

2.8.2 Extrapulmonary tuberculosis
The need for follow-up will depend on factors such as the site of disease and whether the treatment course was uneventful or not. Routine review of patients with lymph node TB and an uncomplicated treatment course is probably unnecessary (patients can be instructed to return if lymphadenopathy recurs), whereas

regular follow-up (including imaging, and liaison with other relevant hospital units) would be indicated for most patients with more complicated forms of TB, such as CNS and bone and joint disease.

2.8.3 Longer follow-up
The following patients may require longer-term follow-up:

- Patients who have not been treated with a regimen that includes both rifampicin and isoniazid
- Patients with silicosis
- Patients who have not been satisfactorily compliant with treatment
- Patients with drug-resistant tuberculosis
- Patients requiring long-term steroid or immunosuppressive therapy
- Patients with HIV.

2.9 When patients develop problems
2.9.1 Drug hypersensitivity reactions

- Hypersensitivity reactions usually occur within the first 8 weeks of treatment. The most common clinical features of hypersensitivity reactions are rash and fever. The rash is usually erythematous and itchy, and may be macular or papular.
- Pyrazinamide often causes initial facial flushing or pruritus. These symptoms are transient and need not be regarded as reactions.
- Generalised reactions also include fever, rigors, headache, myalgia, periorbital swelling, conjunctivitis, generalised lymphadenopathy, hepatosplenomegaly and occasionally transient jaundice. Rarely exfoliative dermatitis or Stevens-Johnson syndrome may occur.
- Hypotension or shock may occur if a large dose of a drug is given after a previous hypersensitivity reaction.

Management of hypersensitivity

- Minor reactions such as a slight itch that do not distress the patient may be self-limiting. Treatment with an antihistamine may be all that is needed without stopping the antituberculous drugs.
- If the reaction is more than trivial, all drug treatment should be stopped and corticosteroids may need to be used.
- For non-severe reactions, when the reaction has subsided attempt to identify the drug responsible by re-introducing the drugs singly, starting with

Table 2.9.1 Recommended sequence of daily challenging doses

	Drug	Day 1	Day 2	Day 3
	isoniazid	50	100	300
then	rifampicin	75	150	450–600 (on 4th day if required)
then	pyrazinamide	250	500	full dose
then	ethambutol	100	400	full dose

a low challenge dose. This is best done in hospital, and referral to Clinical Immunology is advised for all but the most straightforward cases.

- The recommended sequence of daily challenging doses (in mg) in mild to moderate reactions is given in Table 2.9.1. It is not a desensitisation regimen.
- Formal desensitisation is indicated if:
  - the original reaction was severe
  - a reaction occurs with the first challenge dose, as shown above, and it is decided that the drug must be used.
- Desensitisation should not be performed on any patient who had hepatitis, haemolytic anaemia, purpura, nephritis, or ocular toxicities as manifestations of the drug hypersensitivity.
- Desensitisation will require input from Clinical Immunology, but a suggested approach is as follows:
  - When starting to desensitise, it is usually safe to begin with one-tenth of the normal dose. Then the dose is increased by a tenth each day if tolerated.
  - If the patient has a mild reaction to a dose, the same dose (instead of a higher dose) is given next day.
  - If there is no reaction, the dose is to be increased again by a tenth each day.
  - If a reaction is severe (which is unusual), a lower dose is used and then increased more gradually.
  - If a reaction occurs with the second challenge dose, desensitisation can be started with the first challenge dose. Then the dose is increased by the amount equal to the first challenge dose each day.
- A rapid oral desensitisation method for rifampicin and ethambutol was described by Matz et al. (1994).

2.9.2 Gastrointestinal symptoms

Almost any medication can cause upper gastrointestinal (GI) symptoms (anorexia, nausea, vomiting, epigastric discomfort) in susceptible individuals. Among first line antituberculous drugs, rifampicin is the most common cause, although pyrazinamide is responsible in some instances. However, rifampicin is the most important member of combination antituberculous therapy, so every effort should be made to continue this drug.

Because upper GI symptoms may be due to drug-related hepatitis, LFTs must be done in all individuals who present with such symptoms, and all medications should be ceased if transaminases are greater than 3 times the upper limit of normal, as discussed in the following section.

GI symptoms can often be controlled by advising the patient to take rifampicin with food (rather than on an empty stomach), and by use of symptom-relief medications such as H2 blockers. If symptoms persist, or are severe to begin with, antituberculous drugs may need to be discontinued. If unrelated GI disease is suspected, appropriate referral for diagnostic investigation should be arranged.

If medications are ceased then restarted, gradual dose escalation of rifampicin after reintroduction of HZE should be tried. If the GI symptoms are caused by pyrazinamide, it is usually simplest to omit this medication and to extend the duration of treatment to 9 months (see Chapter 1).

2.9.3 Abnormal liver function tests

See also section 2.7.4 Blood tests (liver function tests) and Chapter 5, Diagnosis and management of latent TB infection.

- Modest elevations of hepatic transaminases (AST/ALT) are not uncommon in the pretreatment liver function tests of TB patients.
- Minor transaminase elevations are also common in the early stages of TB treatment, but the risk of clinical hepatitis is much lower (see Table 2.9.3).
- Isoniazid and pyrazinamide are the most common causes of drug-induced hepatitis in patients on TB treatment. The risk of isoniazid hepatotoxicity increases with age.
- As discussed in section 2.7.3, in those whose initial LFTs are normal we do not routinely monitor LFTs unless the patient develops upper gastrointestinal symptoms, is aged 60 or older, has chronic hepatitis B or C or another form of chronic liver disease, is a heavy drinker of alcohol, is HIV-infected, or is pregnant.

Table 2.9.3 Risk of clinical hepatitis with antituberculous drugs

Type of Treatment	Drugs	Number of patients	Percentage with clinical hepatitis
Treatment of latent TB infection	isoniazid alone	38 257	0.6%
Multi-drug regimen	isoniazid-containing regimens without rifampicin	2 053	1.6%
Multi-drug regimen	rifampicin-containing regimens without isoniazid	1 264	1.1%
Multi-drug regimen	rifampicin + isoniazid-containing regimens	6 105	2.55% (US series 3%) (UK series 4%)

- Although the specific cause of hepatitis cannot be determined by the pattern of LFT abnormality in general, if the pattern is hepatocellular with enzymes elevated and out of proportion to bilirubin or alkaline phosphatase, the cause may be isoniazid, rifampicin, or pyrazinamide. Rifampicin is usually implicated if the pattern is cholestatic (elevated bilirubin or alkaline phosphatase out of proportion to enzyme elevations). Very rarely, ethambutol causes hepatitis with a hepatocellular pattern.

Management of abnormal LFTs

VIDS follows the recommendations of the ATS for management of abnormal LFTs during TB therapy (see Figure 2.9.1).

- If the AST/ALT is < 2× the upper limit of normal, liver function should be repeated in 2 weeks:
  - if transaminase levels remain < 2× the upper limit of normal, further repeat tests are only required for symptoms
  - if the repeat test shows an AST/ALT level > 2× the upper limit of normal, management should be as below.
- If the AST/ALT is 2–5× the upper limit of normal, and the patient is asymptomatic, treatment can be continued but liver function should be monitored weekly for 2 weeks then 2-weekly until normal.
- If the AST/ALT level rises to > 5× the upper limit of normal or the bilirubin level rises and AST/ALT is > 3× the upper limit of normal, antituberculous therapy should be stopped.

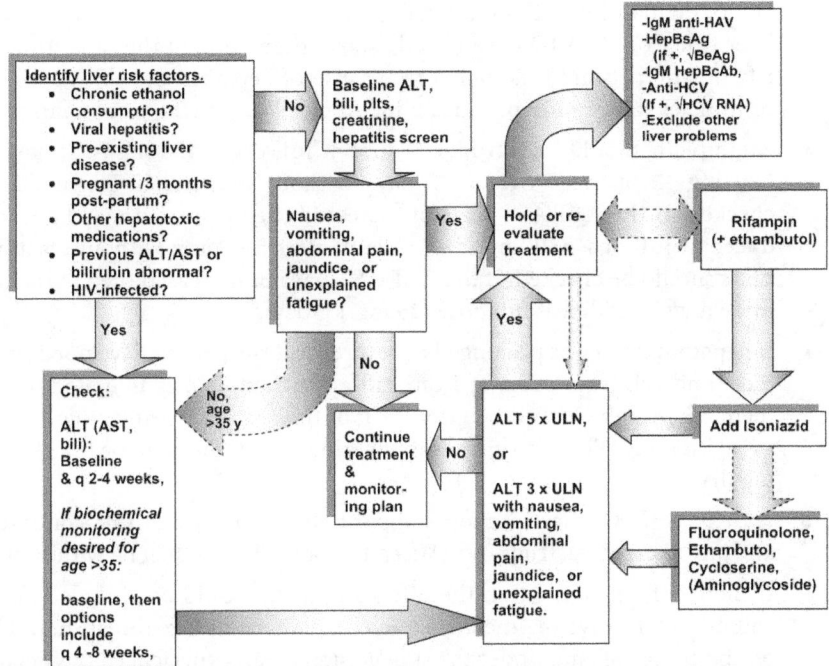

Figure 2.9.1 Monitoring for hepatotoxicity during treatment of TB disease. *Dotted lines* signify management according to physician's discretion. ALT = alanine aminotransferase; AST = aspartate aminotransferase; HCV = hepatitis C virus; HepBsAg = hepatitis B surface antigen. Reprinted with permission of the American Thoracic Society. Copyright © 2012 American Thoracic Society. Saukkonen JJ, Cohn DL, Jasmer RM et al. 2006. Hepatotoxicity of antituberculosis therapy. American Journal of Respiratory and Critical Care Medicine; 174:935–52. Official Journal of the American Thoracic Society.

- If the patient is not unwell and the form of TB is non-infectious, no treatment needs to be given until liver function normalises. This approach requires careful assessment to ensure patients conform to the above description.
- If the individual has extensive pulmonary, or disseminated, TB, or is HIV-positive, institution of a new regimen with less potential for hepatotoxicity (e.g. amikacin, ethambutol, moxifloxacin) may be indicated before liver enzymes normalise. This treatment should be given in hospital. If other reserve drugs are used, any potential hepatotoxicity should be considered.
- Once liver function has returned to < 2× the upper limit of normal (or near to baseline levels if initially abnormal), full dosages of the original drugs can be reintroduced sequentially in the order rifampicin–isonia-

zid–pyrazinamide with monitoring of the patient's clinical condition and liver function. In VIDS, we usually start ethambutol at the same time as rifampicin, because it is such a rare cause of hepatitis and this approach minimises the amount of time on incompletely active TB treatment.
- Rifampicin should be introduced first at full dose (450 mg (< 50 kg) or 600 mg (≥ 50 kg)). After 5–7 days without reaction LFTs should be checked. If the ALT/AST has not increased then isoniazid may be introduced at full dose (300 mg/day). Again, after 5–7 days without reaction, LFTs should be checked. If the ALT/AST has not increased then pyrazinamide may be added at full dose (25 mg/kg/day).
- If hepatotoxicity was prolonged or severe, or if the patient developed jaundice, and reintroduction of isoniazid and rifampicin is tolerated, treatment should be continued for 9 months without reintroduction of pyrazinamide. This also applies if pyrazinamide is found to be the offending drug.
- If there is no further reaction standard chemotherapy can be continued and any alternative drugs introduced temporarily can then be withdrawn.
- If there is further reaction the offending drug should be excluded and a suitable alternative regimen used. Such an alternative regimen should be on the advice of, and under the supervision of, an experienced physician.
- A recent randomised controlled trial of three approaches to reintroduction of TB therapy in the setting of hepatotoxicity suggests that it may be safe to reintroduce rifampicin, isoniazid and pyrazinamide concurrently at full dose. This study compared the above ATS recommended approach with staggered reintroduction of isoniazid then rifampicin then pyrazinamide (as previously practised in VIDS and reflecting the BTS guidelines) and immediate reintroduction. There was no significant difference in the frequency of recurrent drug-induced hepatitis. This was a small ($n = 175$) single-centre study and requires replication and further consideration before the concurrent reintroduction of rifampicin, isoniazid and pyrazinamide can be recommended.

2.9.4 Visual disturbance
- Visual impairment is a rare but serious adverse drug reaction in patients on TB medications. It is usually due to ethambutol, but it has also been reported with isoniazid. While it is exceedingly rare when ethambutol is given at the dose of 15 mg/kg/day, blindness has occurred with this dose (ocular toxicity is dosage related). The incidence of ocular toxicity is increased in those with impaired renal function.

- Visual impairment is due to retrobulbar neuritis. Onset is usually abrupt, but it may be insidious, occurring 3–6 months after commencement of drug. The manifestations are blurring of vision, decreased visual acuity, central scotoma and loss of ability to see green and sometimes red. Colour visual disturbance may be detectable when there is little or no decrease in visual acuity.
- In many cases, if ethambutol is stopped promptly, vision slowly returns. Recovery may take 3–12 months. Sometimes, there is a period of worsening before improvement. Permanent and total visual loss has been reported, even with prompt cessation of ethambutol. In some case series this may occur in up to 50% of cases. Renal function should be checked pre-treatment, and ethambutol should be avoided in patients with impaired renal function.
- The recommended dose and duration should not be exceeded. The maximum dose of ethambutol is 15 mg/kg.
- Any history of eye disease should be recorded in the notes.
- A pre-treatment record of visual acuity should be made by the Snellen test and colour vision documented using Ishihara plates. Ethambutol should not be given to those with pre-existing significant reduced visual acuity, as these patients may not notice further minor deterioration of vision.
- The patient should be told that ethambutol may affect vision and that drugs should be stopped immediately should vision become impaired. To guard against lack of compliance the patient may be assured that this risk is very small.
- This advice should be documented in the medical history, and the family doctor advised.
- Ask patients on ethambutol about symptoms of visual disturbance at each outpatients visit. Patients complaining of visual disturbance during TB treatment should be referred to an ophthalmologist for detailed examination and treatment should be discontinued pending this examination.
- Patients who are continuing on ethambutol beyond 2 months require monthly visual acuity and colour vision measurement. If progressive visual impairment is found, refer the patient immediately for formal ophthalmological assessment. Ethambutol is best avoided in children too young for objective eye tests, and also in adults with language or communication problems that would make assessment difficult.

2.9.5 Arthralgia/arthritis

Arthralgia occurs in some patients on pyrazinamide. It appears to be particularly problematic in African women. Arthralgia usually appears during the first 1–2 months of treatment. Unlike gout, it affects both large and small joints, most commonly shoulders, knees and fingers. Swelling, limitation of movement and tenderness are usually mild. It is commonly self-limiting and responds readily to symptomatic treatment. Sometimes it is necessary to cease pyrazinamide.

Symptoms are not correlated with serum measurements of uric acid, and treatment with allopurinol is thought to be unhelpful.

2.9.6 Acute psychosis

Rifampicin and pyrazinamide are not associated with psychotic side-effects.

Isoniazid, in a standard dose of 5 mg/kg, has been identified as a triggering factor for the onset and relapse of schizophrenia. In healthy individuals psychosis due to isoniazid is very infrequent unless the dose is > 15 mg/kg. Resolution of psychotic symptoms has been reported with cessation of isoniazid; however, antipsychotic medication and involvement of psychiatry services are important.

Isoniazid-induced encephalopathy (manifested with confusion) has occurred in dialysis patients. Observational reports have suggested that the co-administration of high-dose pyridoxine (100 mg daily) in renal patients receiving isoniazid may reduce this complication. Such patients are also prone to pellagra (diarrhoea, dermatitis and dementia) owing to increased requirement and malnutrition-related deficiency of niacin. Isoniazid enhances the risk of pellagra by interfering with conversion of tryptophan to niacin. In the absence of typical dermatitis or diarrhoea, pellagra psychosis has been mistaken as cases of isoniazid-induced encephalopathy.

Neuropsychiatric complications of isoniazid occur more frequently in patients with predisposing factors: old age, diabetes, hepatic insufficiency, alcoholism, malnourishment, hyperthyroidism, slow acetylators, brain damage, past or family history of psychotic illness, high dose of isoniazid, concomitant administration of monoamine oxidase inhibitors.

2.9.7 When lymph nodes enlarge

Tuberculous lymphadenitis under treatment often fluctuates markedly in its clinical course. In 25% of the patients the glands may enlarge or new ones may appear. This paradoxical response may occur after initial improvement, and in 5–10% of the patients it may occur after the completion of treatment. The glands may become attached to the skin and discharge caseous material. Glandular enlargement in the chest may cause airway or superior mediastinal obstruction. In this situation corticosteroid therapy is indicated. Otherwise reassure the patient that this event is common and does not signify failure of treatment. A case-controlled study showed no benefit of corticosteroid therapy for this paradoxical reaction involving tuberculous lymph nodes (Hawkey et al. 2005).

Surgical treatment may occasionally be required for nodes that 'point' where discharge of caseous material is imminent.

Recrudesent or enlarging lymph nodes in patients with tuberculosis is a well recognised phenomenon in HIV-positive patients, particularly in the setting of initiation of antiretroviral therapy when the CD4 cell count is low (see Chapter 6, HIV and tuberculosis).

2.9.8 When haemoptysis occurs

Haemoptysis may occur during treatment and after treatment is completed. It is usually not a sign of disease activity, but sputum specimens should still be sent for AFB smear and culture. If haemoptysis is persistent or recurrent one should exclude a coexistent carcinoma of the lung or bronchiectasis. If haemoptysis is massive patients must be admitted to hospital for observation and management.

2.10 Usual response to treatment

2.10.1 Cough

Cough frequency declines rapidly following treatment - to 35% of initial cough counts within 2 weeks of initiation of treatment.

2.10.2 Fever

About 60% of febrile patients with TB became afebrile within 2 weeks with a mean of 16 days and median of 10 days. The remainder became afebrile more than 2 weeks after commencement of treatment; among this group the average time to defervesce can be as long as 1 month (Kiblawi et al. 1981).

2.10.3 Sputum smear

For drug-sensitive pulmonary TB, the median time to AFB sputum smear conversion is 3–4 weeks, and 80% of patients will be smear negative after 2 months of treatment. The duration of smear positivity is associated with the degree of pretreatment smear positivity; for example, less than 5% of patients with grade 1+ or 2+ smears will still have positive smears after 2 months, compared with up to 50% of patients with initial grade 4+ smears. Smear positivity is also prolonged in patients with cavitary disease.

The two most important factors to consider in patients with persistently positive smears, or in those exhibiting a clearcut fall and rise phenomenon, are emergence of resistance and poor adherence.

The presence of AFB in sputum smears during treatment may occur in up to 20% of patients after sputum culture has become negative. This phenomenon may be seen as early as after 4 weeks of treatment. Beyond 12 weeks of treatment most persistently positive smears (about 2/3) are associated with negative cultures.

2.10.4 Sputum culture
Sputum culture usually becomes negative in patients being treated with both isoniazid and rifampicin:

- at 1 month in > 50%
- at the end of 2 months in 85%
- at the end of 3 months in > 90%

Patients whose sputum has not converted by 3 months of treatment should be carefully re-evaluated.

- Assess compliance. Consider directly observed therapy.
- Check drug susceptibility tests (the results might have been missed).
- Consider the possibility of non-tuberculous mycobacterial infection.
- Do not change therapy unless the patient is clinically ill. It is preferable to wait for results of drug susceptibility tests before altering therapy.
- If treatment is to be changed, never add one new drug at one time. Include at least 2 drugs to which the organisms are susceptible.
- Evaluation should be performed at least monthly until sputum conversion occurs.
- Patients in whom sputum culture results have not converted to negative after 5 to 6 months of therapy are considered treatment failure.

2.10.5 Radiological improvement
New infiltrates may develop within 3–5 weeks after commencement of treatment. It does not mean failure of treatment at this stage. It may be an immunological phenomenon.

CXR usually shows improvement 1 to 3 months after initiation of treatment.

Failure to show improvement after 3 months indicates wrong diagnosis or inadequate treatment.

CXR usually resolves or becomes stable in 90% at 6 months

References and further reading

Fortun J, Martin-Davila P, Molina A, et al. Sputum conversion among patients with pulmonary tuberculosis: are there implications for removal of respiratory isolation? J Antimicrob Chemother, 2007; 59:794-8.

Hawkey C, Yap T, Pereira J, et al. Characterization and management of paradoxical upgrading reactions in HIV-uninfected patients with lymph node tuberculosis. Clin Infect Dis 2005; 40:1368-71.

Horne DJ, Johnson CO, Oren E, Spitters C, Narita M. How soon should patients with smear positive tuberculosis be released from inpatient isolation? Infect Control Hosp Epidemiol 2010; 31:78-84.

Kiblawi SSO, Jay SJ, Stonehill RB, Norton J. Fever response of patients on therapy for pulmonary tuberculosis. Am Rev Resp Dis 1981; 123:20-4.

Management, control and prevention of tuberculosis: guidelines for health care providers 2002-2005. Victorian Government Department of Human Services, Melbourne, 2002.

Matz J, Borish LC, Routes JM, Rosenwasser L. Oral desensitization to rifampin and ethambutol in mycobacterial disease. Am J Respir Crit Care Med 1994; 149:815-17.

Melamud A, Kosmorsky GS, Lee MS. Ocular ethambutol toxicity. Mayo Clin Proc, 2003; 78:1409-11.

Saukkonen JJ, Cohn DL, Jasmer RM, et al., on behalf of the ATS Subcommittee. Hepatotoxicity of antituberculosis therapy. Am J Resp Crit Care Med 2006; 174:935-52.

Sharma SK, Singla R, Sarda P, et al. Safety of 3 different reintroduction regimens of antituberculosis drugs after development of antituberculosis treatment-induced hepatotoxicity. Clin Infect Dis 2010; 50:833

WHO. Treatment of tuberculosis guidelines. 4th edition. World Health Organization, Geneva, 2010.

Yee D, Valiquette C, Pelletier M, et al. Incidence of serious side effects from first-line antituberculosis drugs among patients treated for active tuberculosis. Am J Resp Crit Care Med 2003; 167:1472-7.

Chapter 3

Evaluation and treatment of drug-resistant tuberculosis

3.1	Definition of drug-resistant tuberculosis	37
3.2	Clinical and microbiological assessment of patients with drug-resistant TB	38
3.3	Classifying drugs used to treat drug-resistant TB	39
3.4	Treatment of drug-resistant TB	40
3.5	Monitoring MDR-TB treatment	47
3.6	Post-treatment evaluation	49
3.7	Extensively drug-resistant tuberculosis	49
3.8	Alternative treatments for MDR-TB and XDR-TB	50

3.1 Definition of drug-resistant tuberculosis

Drug-resistant tuberculosis refers to a strain of *M. tuberculosis* that is resistant to at least one antituberculous drug. Multidrug-resistant tuberculosis (MDR-TB) strains are resistant to at least isoniazid and rifampicin. Extensively drug-resistant tuberculosis (XDR-TB) has recently been described and denotes MDR-TB that is also resistant to a fluoroquinolone agent and one or more second-line injectable agents (amikacin, capreomycin or kanamycin). In this chapter, unless otherwise specified, drug-resistant TB refers to single, multi- and extensively drug-resistant TB, but the principal focus of this chapter will be MDR-TB.

In Australia, isoniazid (with or without streptomycin) resistance occurs in 7–10% of TB isolates. Local rates of MDR-TB have remained relatively stable at 0.5–2% for the past 10–15 years, but the global rate of MDR-TB is now estimated to be 3.6%. Since almost all TB in Australia is imported, it is virtually inevitable that we will see an increase in MDR-TB cases in the coming years. Extensively drug-resistant tuberculosis has not been reported in a previously untreated patient in Australia but is potentially only a jumbo jet flight away.

3.2 Clinical and microbiological assessment of patients with drug-resistant TB
3.2.1 Risk factors
Clinical features at presentation do not distinguish patients with drug-sensitive from those with drug-resistant TB. However, the following features (some identified in overseas studies and not necessarily applicable in Australia) are risk factors for MDR-TB, and if present may prompt consideration of empirical use of second-line antituberculous medications in addition to the usual 4-drug regimen in some circumstances (discussed in more detail in section 3.4).

Epidemiological clues
- Close contact with an individual with MDR-TB.
- Birth in a country with a high prevalence of MDR-TB.
- Previous treatment for tuberculosis, in particular if treatment was incomplete.
- Previous hospitalisation in a hospital with an outbreak of a drug-resistant strain of TB, particularly if housed on the ward where the outbreak occurred.

Markers of non-response to standard TB treatment
- Failure to defervesce after 2 weeks of treatment with a standard 4-drug regimen; however, persistent fever can also be caused by severe miliary disease or another concomitant infection, so this is not a specific sign of MDR-TB.
- Persistent symptoms despite treatment, and detection of bacilli in sputum specimens.
- Persistently positive culture after 3 months of initial treatment with isoniazid, rifampicin, pyrazinamide, and ethambutol – for fully drug-sensitive TB, 80 to 90% of patients with positive pre-treatment sputum cultures should have converted to negative after 2 months of treatment, and nearly 100% should have negative cultures within 5 to 6 months.
- Reversion to a positive culture after having initially converted cultures from positive to negative.

In practice, the most useful epidemiological pointer to the presence of MDR-TB is the combination of previous treatment for TB and birth in a country with a high TB prevalence. The WHO undertakes ongoing global surveillance of TB drug resistance in selected countries and regions, and uses this data to provide estimates of the rates of MDR-TB in newly diagnosed and re-treatment cases by country. For example, in the WHO's latest report (2010) the estimated rates of MDR-TB in China are 5.7% for newly diagnosed and 25.6% for re-treatment cases. The latter figure is certainly high enough to warrant consideration of empirical MDR-TB treatment prior to results of susceptibility testing becoming available, and PCR testing for rifampicin resistance should be requested on clinical specimens (especially if AFB smear positive) or on primary cultures (see following section).

3.2.2 Sputum microscopy, culture and nucleic acid amplification testing

Clinical specimens are essential for confirmation of MDR-TB and for drug susceptibility testing (DST) and to document culture conversion.

The availability of nucleic acid amplification tests (NAAT) for rapid detection of multi-drug resistance can dramatically shorten the time to diagnosis and allow earlier institution of appropriate treatment. Assays such as Genotype MTBDR-Plus and GeneXpert MTB/RIF are highly sensitive and specific for detection of rifampicin resistance in AFB smear-positive sputum specimens (sensitivity is lower in AFB smear-negative specimens), and give results in 1–2 days. These assays are being rolled out by WHO in high MDR-TB prevalence countries. In Victoria, the Mycobacterial Reference Laboratory offers molecular resistance testing, though not currently on a routine basis. We expect that the availability and use of these tests will increase in the coming years.

The presence of dead bacilli can produce 'false positive' smear results early in treatment and a similar phenomenon has been reported with NAATs. Up to 20% of patients (most commonly those with cavitary disease) who are initially smear- and culture-positive may continue to have positive smears after cultures become negative for up to 20 weeks after commencement of treatment.

3.3 Classifying drugs used to treat drug-resistant TB

For the purposes of treatment of drug-resistant tuberculosis (predominantly MDR-TB), WHO classifies antituberculous drugs into a hierarchy of 5 groups:

- Group 1 – first-line oral agents (isoniazid, rifampicin, ethambutol and pyrazinamide).
- Group 2 – injectable agents: aminoglycosides (streptomycin, kanamycin and amikacin) and cyclic polypeptides (capreomycin).
- Group 3 – fluoroquinolones (potency based on descending order of *in vitro* activity: moxifloxacin > levofloxacin > ciprofloxacin = ofloxacin).

- Group 4 – oral bacteriostatic second-line agents: thioamides (e.g. prothionamide and ethionamide); serine analogues (e.g. cycloserine and terizidone); salicylic acid derivatives (e.g. para-aminosalicylic acid – PAS); thioacetazone.
- Group 5 – agents with *in vitro* activity but with unclear efficacy ('third-line agents'): clofazimine, amoxycillin–clavulanic acid, clarithromycin, linezolid (other authorities also include imipenem and high dose isoniazid in this group, and possibly cotrimoxazole).

In Australia, the most widely used drugs in groups 2, 3 and 4 are amikacin, moxifloxacin, prothionamide (not ethionamide), cycloserine and PAS. In Victoria, these drugs are now funded by DoH; clinicians are required to complete a DoH form (available from pharmacy) certifying that pre-specified criteria have been met. Prothionamide and PAS have to be prescribed through the Special Access Scheme: liaise with pharmacy about specific details.

Table 3.3.1 lists dosing and other information about the agents most commonly used for MDR-TB treatment.

3.4 Treatment of drug-resistant TB
3.4.1 General principles
- Expert consultation must be sought for individuals with confirmed or suspected drug-resistant TB, especially MDR-TB and XDR-TB. Registrars should always discuss issues such as design of an initial regimen or management of drug toxicity with a consultant experienced in the management of drug-resistant TB.
- Use of standardised protocols for treatment of known or suspected MDR-TB and XDR-TB is often not possible because of factors such as the drug susceptibility results of an individual isolate, patient co-morbidities, and drug toxicity.
- Patients with known or suspected pulmonary MDR-TB or XDR-TB should be hospitalised for isolation. Hospitalisation of other MDR-TB or XDR-TB patients is usually necessary in order to ensure compliance and to monitor drug intolerance and toxicity, drug levels and response.
- Treat MDR-TB patients under directly observed therapy (DOT). Arrangements will need to be discussed and post-discharge care coordinated with the public health TB program.
- Intermittent therapy regimens should not be used for patients with MDR-TB or XDR-TB.

Table 3.3.1 Properties of selected antituberculous drugs used in MDR-TB treatment

Drug	Type of activity	Usual daily dose	Peak serum concentration (mg/L)	Adverse effects
Amikacin	Bactericidal	15 mg/kg IM or IV	25–35	Hearing loss, vestibular toxicity, renal toxicity
Capreomycin	Bactericidal	15 mg/kg IM or IV	25–35	As for amikacin
Ethionamide/ prothionamide	Bactericidal	250 mg PO bd or tds (administer with pyridoxine 100 mg)	1–5	Nausea, vomiting, metallic taste, hepatitis, hypothyroidism, CNS effects
Moxifloxacin	Bactericidal	400 mg daily (PO or IV)	3–4	GI upset, dizziness, headaches, rarely other CNS effects and tendon rupture
Cycloserine	Bacteriostatic	250 mg PO bd or tds (administer with pyridoxine 50–100 mg)	20–35	CNS effects: dizziness, psychoses, seizures
Para-aminosalicylic acid	Bacteriostatic	4 g PO bd or tds	20–60	GI upset, hepatitis, hypokalaemia, hypothyroidism

- Effective MDR-TB treatment requires the use of at least four active drugs (preferably five drugs in the initial phase) and prolonged therapy (for at least 18 months).

3.4.2 Selection of empiric regimen (susceptibilities pending)
Non-MDR-TB

Patients suspected on epidemiological grounds to have resistance to INH can be started on standard therapy with HRZE.

MDR-TB

If MDR-TB is anticipated or considered likely but results of susceptibility testing are not available, options are:

- Start HRZE and await results of molecular and conventional drug susceptibility testing
- Start empiric regimen with MDR-TB activity, comprising HRZE plus moxifloxacin plus amikacin plus prothionamide. This will provide good activity against MDR-TB, except for the rare strains that are resistant to all first-line agents or for XDR-TB.

Because of the pill burden and risk of toxicity, the second approach is only indicated in those with epidemiological risk factors for MDR-TB who are critically ill or whose condition is deteriorating on HRZE.

3.4.3 Specific treatment regimens according to resistance patterns

Perceived inconsistencies in recommended drug treatment regimens for drug-resistant TB arise from:

- Differing expert opinions in the management of drug-resistant TB, in particular MDR-TB, which are influenced by personal experience
- Lack of rigorous randomised controlled trials in the treatment of MDR-TB
- Technical difficulties in performing *in vitro* DST, especially for second-line agents. Interpretation and application of DST requires experienced staff, local knowledge of drug usage, and appreciation of significant cross-resistance patterns of TB strains. For example, susceptible and resistance results for isoniazid and rifampicin are reliable, while 'susceptible' results are more reliable than 'resistant' results for streptomycin and ethambutol. Rates of cross-resistance are 15–20% between isoniazid and ethionamide, 20–60% between kanamycin and capreomycin, and 100% between amikacin and kanamycin.

The following are guidelines for the treatment of TB strains with specific resistance patterns. For MDR-TB options are seldom clear-cut, as individuals often will have already received trials of some of the medications.

Non-MDR-TB

Management of these infections is usually straightforward. Treatment duration is generally longer than for drug-sensitive TB, but most patients can be treated with effective, well-tolerated regimens.

Isoniazid resistance (± streptomycin)

Resistance found after initiation of treatment with standard 4-drug regimen. Options are:

2 HRZE/ 4 RZE (US guidelines)

2 HRZE/ 7 RZE (VIDS preferred, if pyrazinamide tolerated)

2 HRZE/ 10 RE (UK guidelines)

Resistance known before initiation of treatment. Options are:

6 RZE (US guidelines)

9 RZE (VIDS preferred, if pyrazinamide tolerated)

2 RZE + injectable aminoglycoside/ 7 RE (UK guidelines)

2 RZE/ 10 RE

A fluoroquinolone may be added if there is extensive disease (US guidelines) but otherwise should not be part of standard management of these infections.

If the patient is HIV-positive or immunocompromised, a 9 or 12 month regimen is advisable, but this is not evidence-based.

In earlier studies where resistance testing was only performed following completion of treatment, isolated baseline INH resistance was shown to have minimal impact on cure rates with the standard 2HRZE/4HR regimen (95%, versus 98% with no resistance), but in developed countries where susceptibility testing is performed routinely, the finding of INH resistance should lead to a change to one of the regimens above.

The use of isoniazid in the treatment of isoniazid-resistant TB is controversial. Large populations of isoniazid-resistant organisms may contain some isoniazid-susceptible organisms. Some feel that these susceptible organisms may be more virulent than the resistant bacilli, and continue isoniazid even when the laboratory has reported isoniazid resistance. However, if other effective medications are included in the regimen, the addition of isoniazid has not been shown to improve the effectiveness of the regimen.

Isoniazid and ethambutol resistance
If initial standard treatment phase:

2 HRZE/ 7–10 RZQ[1] ± IA[2]

If no initial treatment phase:

9–12 RZQ ± IA

Increasing experience with fluoroquinolones indicates they have similar activity to isoniazid against *M. tuberculosis*, hence use of an injectable agent is optional for this resistance pattern.

Rifampicin resistance
Options are:

2 H(R)ZE/ 16 HE (some US authorities recommend addition of fluoroquinolone, in which case duration of treatment could be shortened to 12 months)

2 H(R)ZEIA/ 10 HE

9 HZIA

Use of one of the regimens that includes an injectable agent is advised for patients with extensive disease.

MDR-TB

Management of these infections is challenging for doctor and patient alike. Treatment involves prolonged use of drugs that are unfamiliar to most clinicians, are not readily available, may be difficult to administer, and are often poorly tolerated. The treatment course must be supervised by an experienced physician, and close liaison with public health nurses is essential.

Choose drugs based on the hierarchical order of potency. Start with any remaining Group 1 drugs to which the isolate is susceptible, then include one agent from each of Group 2 and Group 3, and if necessary use one or more agents from Groups 4 and 5 to achieve the aim of using at least four active agents.

- Group 2 – amikacin is preferred because it is more readily available than kanamycin and capreomycin and MDR-TB strains are almost always

---

[1] Q = fluoroquinolone – moxifloxacin advised
[2] IA = injectable aminoglycoside

resistant to streptomycin. Amikacin-resistant isolates are always resistant to kanamycin, and should be treated with capreomycin. If possible, use amikacin in combination with pyrazinamide because the combination has good bactericidal activity. Give amikacin IV, usually via a peripherally inserted central catheter (PICC) because prolonged therapy is required.

- Group 3 – moxifloxacin is the preferred fluoroquinolone agent.
- Group 4 – prothionamide is the most widely used agent, PAS and cycloserine less so. Adverse reactions are frequent with these agents so start with small doses of each of these drugs and increase to full doses (referred to as 'ramping') over 5–10 days.

Treatment duration
Treatment is divided into the intensive phase (when the injectable agent is given) and the continuation phase. The duration of the intensive phase is a minimum of 6 months or for 6 months following culture conversion, and the total duration of treatment should be for a minimum of 18 months after culture conversion. Treatment should be extended in cases of highly resistant infections, extensive disease, delayed microbiological response and (probably) coinfection with HIV.

The role of thoracic surgery
Surgery should only be undertaken in patients able to tolerate the procedure (a lobectomy or pneumonectomy) who have focal, usually cavitary, disease. General indications are:

- Positive cultures beyond 4 to 6 months of treatment for MDR-TB

or

- Extensive patterns of drug resistance that are unlikely to be cured with chemotherapy alone.

Surgery should be performed by an experienced surgeon.

Isoniazid and rifampicin resistance
Intensive phase: ZEQIA + Pro[1]
Continuation phase: ZEQ + Pro

Prothionamide may be omitted in selected cases of 'low burden' disease, especially if the patient has already responded well to standard four-drug therapy, e.g. isolated lymph node TB; pleural TB; smear negative, non-cavitary, localised pulmonary TB.

---

1   Pro = prothionamide

Isoniazid and rifampicin and ethambutol resistance
Intensive phase: ZQIA + Pro (or PAS[1]) + Cs[2]
Continuation phase: ZQ + Pro (or PAS) + Cs

The combination of prothionamide plus PAS is associated with a high incidence of gastrointestinal side effects and is best avoided unless cycloserine cannot be given.

Isoniazid and rifampicin and pyrazinamide resistance
Intensive phase: EQIA + Pro (or PAS) + Cs
Continuation phase: EQ + Pro (or PAS) + Cs

The combination of prothionamide plus PAS is associated with a high incidence of gastrointestinal side-effects and is best avoided unless cycloserine cannot be given.

The dose of ethambutol may be increased to 25 mg/kg but retrobulbar neuritis is more common at this dose, and formal ophthamological monitoring is necessary.

Isoniazid and rifampicin and pyrazinamide and ethambutol resistance
Intensive phase: QIA + Pro + Cs + PAS
Continuation phase: Q + Pro + Cs + PAS

- Extend the intensive phase to 9–12 months if possible.
- Addition of a Group 5 (third-line) agent may be necessary.
- Treatment should be continued for at least 24 months after culture conversion.
- The combination of prothionamide and PAS *is* recommended in this situation because of limited treatment options.
- The threshold for surgery should be lower than with lesser degrees of resistance.

Isoniazid and rifampicin and pyrazinamide and ethambutol and amikacin resistance
Intensive phase: Q + Capreomycin + Pro + Cs + PAS ± one or more third-line agents
Continuation phase: Q + Pro + Cs + PAS ± one or more third-line agents

1 PAS = para-aminosalicylic acid
2 Cs = cycloserine

- Give capreomycin if susceptible (there is some cross-resistance between amikacin and capreomycin).
- Extend the intensive phase to 9–12 months if possible.
- Treatment should be continued for at least 24 months after culture conversion.
- The combination of prothionamide and PAS *is* recommended in this situation because of limited treatment options.
- Surgery should be considered at an early stage because of the high level of resistance, especially if capreomycin cannot be given.

3.5 Monitoring MDR-TB treatment

3.5.1 Bacteriological status

It is mandatory to monitor bacteriological (smear and culture) status monthly from the second month until the sixth month, and then quarterly until the end of treatment. In contrast to drug-sensitive TB, patients with MDR-TB who are unable to expectorate sputum *should* undergo sputum induction to document conversion to culture negativity.

For a patient with a positive TB culture 3 months after starting treatment, there are two possible approaches.

1. If not acutely ill, and clinically stable, maintain on the 'holding regimen' until the new drug susceptibility results become available.
2. If acutely ill, or clinically deteriorating, add at least two (preferably three) new medications, based on an assessment of what remaining medications the organism is likely to be susceptible to. The original medications should be continued pending repeat drug susceptibility testing results. Never add a single antituberculosis medication to a failing regimen.

3.5.2 Drug toxicity

Close monitoring for drug toxicity of injectable and second-line agents is essential.

- Aminoglycoside toxicity:
    - warn (and subsequently enquire) about balance or coordination problems, hearing difficulty or tinnitus
    - check renal function and electrolytes weekly for the first 2 weeks then 2–4 weekly thereafter, and calcium and magnesium monthly

- check trough amikacin levels weekly for first 2 weeks, then monthly, or at any time if significant changes occur in renal function
- perform baseline audiometry, and subsequently as advised by audiologist (usually 2-monthly)
- at baseline and at each follow-up visit, perform simple vestibular function testing (Rhomberg's, doll's eye manouevre, dynamic visual acuity) and refer for formal testing if any abnormalities are detected.
- Liver function tests – check 1–2 weekly for the first 4 weeks then monthly.
- Thyroid function tests – check 3-monthly for patients on PAS or prothionamide.

If the patient is experiencing a severe side-effect due to a specific drug that precludes its further use, such as ototoxicity from an aminoglycoside or gout from pyrazinamide, but the regimen is not failing (i.e. the patient has improved clinically and cultures have converted from positive to negative) and it is too soon to discontinue any medications, options are:

- Cease the medication responsible for the side-effect and continue on the remainder of the antituberculous treatment regimen
- Substitute with a new, previously unused agent. This does not risk the emergence of resistance since the prior antituberculous regimen was not failing.

If the cause for the adverse reaction (e.g. hepatotoxicity, skin rash) cannot be identified readily, all medications should be discontinued and retested by reintroduction singly into a regimen. In some instances of severe toxicity, hospitalisation for rechallenge with multiple drugs may be needed.

3.5.3 Therapeutic drug monitoring

Routine therapeutic drug monitoring (TDM) is generally not required, with the exception of measurement of aminoglycosides. Few Australian laboratories are able to measure TB drug levels (apart from aminoglycosides, and to a limited extent rifampicin). Although TDM has its proponents its value is uncertain, primarily because there is lack of sufficient data to correlate *in vitro* drug therapeutic ranges with clinical efficacy and response.

Peak serum levels of agents used for MDR-TB treatment are listed in Table 3.3.1. Serum TDM for drugs other than aminoglycosides should be considered if:

- MDR-TB or XDR-TB patients develop symptoms that may be drug-related (e.g. CNS symptoms on cycloserine) and options for alternative agents are limited or non-existent

- The patient has a medical condition, such as malabsorption syndrome (chronic diarrhoea, short gut) or renal inpairment that affects pharmacokinetics and increases risk of treatment failure or toxicity.

3.6 Post-treatment evaluation

Patients who have completed treatment for MDR-TB should be followed up every 6 months for at least 2 years. Clinical review includes symptom review, sputum examination for AFB smear, culture and DST, and CXR (if pulmonary TB).

3.7 Extensively drug-resistant tuberculosis (XDR-TB)

Extensively drug-resistant TB (MDR-TB, plus resistance to a fluoroquinolone agent and to one of the injectable agents amikacin, kanamycin or capreomycin) is an emerging problem. It was first described in HIV-positive patients in KwaZulu-Natal in South Africa, where it was associated with person-to-person transmission and a high mortality. Approximately 44% of MDR-TB isolates have resistance to at least one second-line drug, and 9.9% have resistance to at least three second-line drugs. Nearly 50% of XDR-TB isolates are also resistant to all first-line agents. Thus, XDR-TB may be resistant to at least seven TB drugs. This renders XDR-TB potentially untreatable because it may not be possible to construct a treatment regimen that consists of more than four drugs to which the organism is susceptible.

Countries and regions with high rates of XDR-TB include the Baltic states and some other countries of the former Soviet Socialist Republic (up to 15% of all MDR-TB tested, and up to 20% in some cities and districts), South Africa (10.5%) and China (7.2%). Data are unavailable for most high burden MDR-TB countries because of limited second-line drug susceptibility testing. Rates and numbers of cases of XDR-TB in most industrialised nations are low.

Treatment of XDR-TB should only be undertaken by a physician experienced in the management of drug-resistant TB and in close liaison with the state or territory mycobacterial reference laboratory and public health TB program.

The success rate of XDR-TB treatment varies from 57–66%, where cure is defined as > 5 consecutive culture-negative specimens during the final 12 months of treatment or completion of treatment regimen. Information about results of treatment in the longer term is limited.

The choice of antituberculous drugs follows the general principles of MDR-TB treatment and is based on results of drug sensitivity tests and prior treatment history; at least four active drugs should be used. As with cases of highly drug-resistant MDR-TB, use of Group 5 (third-line) drugs such as linezolid, clofazimine and amoxycillin–clauvulanic acid and consideration of surgical resection may be necessary. Treatment duration is for at least 24 months after culture conversion.

3.8 Alternative treatments for MDR-TB and XDR-TB

High-dose isoniazid
Strains of *M. tuberculosis* identified in the laboratory as isoniazid-resistant often contain mixtures of susceptible and resistant organisms. The use of high-dose isoniazid 16 to 20 mg/kg (i.e. 1–1.5 g/day) would eliminate susceptible organisms and those with low-level resistance. However, high-dose isoniazid is associated with hepatotoxicity, peripheral neuropathy and convulsions, and although the latter two adverse effects are preventable by large doses of pyridoxine, the role of this treatment remains undefined.

Rifabutin
Some rifampicin-resistant strains test susceptible to rifabutin, but as the MICs are just at or below the 'breakpoint' most authorities do not recommend use of rifabutin against these strains.

Clarithromycin
The drug has demonstrated poor *in vitro* activity against *M. tuberculosis* (MICs are generally above 16 mg/L) and (in contrast to its role in the treatment of *M. avium* complex infections; see Chapter 7) it has minimal value in the treatment of MDR-TB.

Clofazimine
There have been anecdotal reports of successful treatment of MDR-TB with regimens containing clofazimine; it may have some use when choices are limited, such as in the treatment for XDR-TB.

Cotrimoxazole
Some MDR-TB strains are susceptible *in vitro* using MIC cut-offs for conventional bacteria, but clinical data are limited to case reports.

Amoxycillin–clavulanic acid
Beta lactam antibiotics penetrate the mycobacterial cell wall and in combination with a beta-lactamase inhibitor have proven bactericidal activity against *M. tuberculosis*. Clinical experience is limited, but drugs such as amoxycillin–

clavulanic acid may have a role in treatment of highly drug-resistant MDR-TB or XDR-TB.

Linezolid
Linezolid has *in vitro* activity against *M. tuberculosis*, including MDR-TB and XDR-TB, and culture conversion has been documented in patients with MDR-TB and XDR-TB treated with linezolid. Significant side-effects, including bone marrow depression and irreversible peripheral neuropathy, often limit this drug to short-term use only.

Thioridazine
Thioridazine is a phenothiazine derivative, a class of neuroleptic drugs used for the treatment of psychosis. Thioridazine has *in vitro* antituberculous activity, but there is no reported clinical experience and phenothiazine side-effects are likely to be problematic.

Investigational agents
New nitroimidazoles (e.g. PA824) have been developed that have very promising *in vitro* and animal model activity against *M. tuberculosis* and have undergone phase I studies and are soon to begin phase II testing. Diarylquinolines (e.g. TMC207) are a novel drug class with potent activity against *M. tuberculosis in vitro* and in a murine animal model. Phase I studies of TMC207 have demonstrated good tolerability and promising antituberculous activity, and a randomised phase II study in patients with MDR-TB is in progress.

References and further reading

Centers for Disease Control and Prevention. Treatment of tuberculosis. American Thoracic Society, CDC and Infectious Diseases Society of America. MMWR Recommendations and Reports. 2003; Vol 52 No. RR-11.

Cole SC, Riccardi G. New tuberculosis drugs on the horizon. Curr Opin Microbiol 2011; 14:570–76.

Francis J. Curry National Tuberculosis Center and California Department of Public Health, 2008: Drug-Resistant Tuberculosis: A Survival Guide for Clinicians, 2nd edn. Available at: <http://www.currytbcenter.ucsf.edu/drtb>.

Gandhi N, Nunn P, Dheda K et al. Multidrug-resistant and extensively drug-resistant tuberculosis: a threat to global control of tuberculosis. Lancet 2010; 375:1830–43.

National Institute for Health and Clinical Excellence. Tuberculosis: Clinical diagnosis and management of tuberculosis, and measures for its prevention and control. Available at: <www.nice.org.uk/nicemedia/pdf/CG033FullGuideline.pdf>.

Shah N, Wright A, Bai G-H et al. Worldwide emergence of extensively drug-resistant tuberculosis. Emerg Infect Dis 2007; 13(3):380–7.

WHO. Multidrug and extensively drug-resistant TB (M/XDR-TB): 2010 global report on surveillance and response. World Health Organization, Geneva, 2010.

WHO. Guidelines for the programmatic management of drug-resistant tuberculosis. Emergency Update 2008. World Health Organization, Geneva, 2008.

Chapter 4

Tuberculosis, pregnancy and perinatal management

4.1	Impact of pregnancy on TB	53
4.2	Impact of TB on pregnancy	53
4.3	Antituberculous drugs in pregnancy	54
4.4	Breast feeding and antituberculous drugs	56
4.5	Management of the newborn after delivery	56
4.6	Screening for tuberculosis during pregnancy	61
4.7	Latent TB infection (LTBI) in pregnancy	61
4.8	Infertility related to tuberculosis	62

4.1 Impact of pregnancy on TB

Pregnancy has no adverse impact on TB if there is no delay in diagnosis. Birth, puerperium and lactation also have no effect on TB or response to treatment. However, the diagnosis of TB may be delayed in pregnancy because pregnant patients with pulmonary TB are more likely to be asymptomatic at time of diagnosis compared with non-pregnant women with pulmonary TB. They are more likely to have non-specific symptoms and to experience a delay in obtaining a CXR than non-pregnant women with TB. The clinical manifestations of pulmonary TB, if present, are the same as those in non-pregnant women.

The likelihood of extrapulmonary TB is more related to the demographic profile of the patient than to pregnancy itself. In Australia, the incidence of extrapulmonary TB is greater among immigrants than those born in this country. The symptoms of extrapulmonary TB are frequently non-specific and may be mistakenly attributed to physiological changes of pregnancy. A high index of suspicion is needed when pregnant immigrants develop symptoms.

4.2 Impact of TB on pregnancy

Maternal and fetal outcome in pregnancy vary with the site of the TB and the timing of diagnosis in relation to delivery. There seems to be no adverse outcome with lymph node TB. In other sites, the outcome of pregnancy is almost always unaffected provided TB is diagnosed early and treated, the exception being the rare instances of congenital TB. In contrast, obstetric and perinatal morbidity is increased in patients with a late diagnosis of pulmonary TB and extrapulmonary TB involving sites other than lymph nodes.

For pulmonary TB diagnosed late in pregnancy, there is an increased incidence of pre-eclampsia, vaginal bleeding, early fetal death, prematurity, small-for-date, low birthweight, and low APGAR scores and perinatal death compared with pregnant women without TB. For extrapulmonary TB in sites other than lymph nodes, late diagnosis can lead to increased antenatal hospitalisation, poor APGAR scores and low birthweight infants compared with healthy controls. Untreated TB represents a far greater hazard to a pregnant woman and her fetus than does treatment of her disease.

Congenital infection may occur by transplacental spread or by aspiration or ingestion of infected amniotic fluid *in utero* or infected genital secretion during birth. These routes of infection are extremely rare. Most cases of neonatal TB occur as a result of airborne spread after delivery.

4.3 Antituberculous drugs in pregnancy

Treatment-related complications are reported but are less common than the reported rates of TB-associated complications as shown in Table 4.3.1 (adapted from Bothamly 2001). The oft-quoted increased risk of isoniazid-associated hepatitis during pregnancy and the post-natal period is the consequence of a single, retrospective and non-statistically significant study.

Isoniazid and **ethambutol** are both category A drugs, and are safe in pregnancy.

Rifampicin (category C)
Current consensus is that rifampicin is not teratogenic and that any risk to the fetus is small. Pregnancy is not a contraindication to use of rifampicin. Rifampicin can give rise to a haemorrhagic tendency in the newborn baby when administered late in pregnancy. Some authorities therefore advise supplemental vitamin K (10 mg/day) for the last 4–8 weeks of pregnancy.

Streptomycin (category D)
Streptomycin is contraindicated in pregnancy as it is ototoxic to the fetus.

Pyrazinamide (category B2)
Pyrazinamide is little studied in pregnancy but is recommended by WHO and is routinely used in pregnancy at VIDS for all cases of TB.

Table 4.3.1 TB-related complications in pregnancy: rates per 100 000 pregnancies

Risk	No TB	TB	Therapy related
Low birth weight	16 500	34 200	
Premature	11 100	22 800	
Small for date	7 900	20 200	
Preeclampsia	4 700	7 400	
Perinatal death	1 600	10 100	
Fetal death	230	2 010	
INH hepatitis			1 600
All drug-related hepatitis			2 700
Fatal hepatitis			9.4–14

Pyrazinamide is strongly recommended in three situations:

1. When multidrug-resistance is suspected.
2. When the pregnant woman is HIV-infected.
3. In a pregnant woman with tuberculous meningitis, especially when isoniazid resistance is a possibility.

Pyridoxine
Supplement in pregnancy should be at a dose of 50 mg/day (instead of 25 mg/day).

4.3.1 Agents used for multidrug-resistant TB (MDR-TB)

The treatment of MDR-TB is more complicated and difficult even in the absence of pregnancy, and overall care should always be under the direction of an individual with expertise in the management of patients with MDR-TB. For further details, refer to Chapter 3, Evaluation and treatment of drug-resistant tuberculosis.

Animal studies have observed teratogenicity with capreomycin, ethionamide and aminoglycosides, and patients who are taking these agents should be counselled to avoid pregnancy. The possibility of teratogenicity should be emphasised and the patient should be advised to use two contraceptive methods. In the event of diagnosis of MDR-TB during pregnancy, expert advice should be sought regarding planning therapy for optimal maternal and fetal outcomes.

4.3.2 Coinfection with HIV and TB
Coinfection with HIV and tuberculosis renders treatment regimens more problematic because of potential drug interactions and increase in frequency of treatment-related side-effects. For further details, see Chapter 6, HIV and tuberculosis.

4.4 Breastfeeding and antituberculous drugs
TB drugs are excreted into breast milk. It is estimated that breastfeeding infants receive no more than 20% of the usual therapeutic dose of isoniazid for infants, and less than 11% of other antituberculous drugs. Potential toxic effects of drugs delivered in breast milk have not been reported.

Australian antibiotic guidelines recommend that breastfeeding infants of mothers taking isoniazid be given pyridoxine 5 mg daily (which must be crushed, and made into a suspension with water).

One way to minimise the level of antituberculous drugs in the baby is for the mother to take her antituberculous drugs immediately after feeding her child and substitute a bottle for the next feed, then go back to her usual feeding pattern till the next day.

4.5 Management of the newborn after delivery
Neonates exposed to TB during or after pregnancy should always be under the care of a paediatrician, but clinicians who look after adult TB patients need to be aware of the principles of management of exposed neonates.

The possibility of neonatal exposure may only become apparent in the early postpartum period when the mother is referred to an adult clinic with possible disseminated TB or smear-positive pulmonary TB. In such a case, it is essential that the treating physician recognises the potential for neonatal infection and organises immediate assessment of the baby at a major paediatric centre (for example in Melbourne at the Royal Children's Hospital or Monash Medical Centre). Subsequently, clinicians treating the mother and baby should maintain close contact, in conjunction with the public health TB program.

4.5.1 Withold BCG if HIV status is unknown
HIV status of the mother should be determined in all cases if not already known. If the mother is HIV-infected then HIV status of the newborn needs to be determined and appropriate HIV-related management initiated. Do not give BCG if the infant is HIV-infected.

4.5.2 Management differs according to the routes and timing of exposure
Scenario 1: risk of haematogenous spread
Mother's TB is likely to be associated with haematogenous spread (e.g. miliary TB, tuberculous meningitis during pregnancy or puerperium) or mother has active pelvic/genital TB during that time.

- The infant has a definite risk of having or developing congenital TB. The onset of congenital TB averages 2–4 weeks (range a few days to a few months) after birth.
- Keep the placenta and send for microscopy and culture, PCR and histological examination.
- At birth, assess neonate for clinical evidence of congenital TB (e.g. fever, respiratory distress, hepatospenomegaly), and do CXR (clinical abnormalities are usually non-specific) and gastric washings for smear and culture. Culture is positive for *M. tuberculosis* in up to 80% of cases and provides important additional information of drug susceptibility. However, gastric washings are rarely smear-positive for acid-fast bacilli and therefore are not usually helpful in early assessment to confirm or exclude the diagnosis of TB. Lumbar puncture to exclude TB meningitis is indicated if abnormalities suggestive of congenital TB are found.
- Investigations should be done to detect other non-specific findings that can be found in congenital TB, including anaemia, altered liver function, raised inflammatory markers, and thrombocytopenia.
- Treat with multiple drug therapy if any evidence of congenital TB is found.
- In the absence of active disease:
    - commence isoniazid (15 mg/kg/day) at birth
    - perform careful clinical assessment frequently during the first 6 months
    - perform a TST (tuberculin skin test) at 4–6 weeks after birth (usually negative during the first few weeks), and repeat the TST at 12 weeks and 6 months; repeat the CXR at 4–6 weeks
    - if the TST and CXR remain negative, continue isoniazid for 6 months
    - if the TST is greater than 5 mm and there is no clinical or X-ray evidence of TB, extend treatment to 9 months
    - perform an interferon gamma release assay (e.g. QFN Gold In-Tube – QFN-GIT) on blood from the baby at birth and at follow-up as it may detect infection sooner and more reliably than TST.

- If there is any doubt about the diagnosis of congenital TB disease, have a low threshold for treating with a full course of antituberculous therapy.
- Give BCG to the infant once congenital TB has been excluded or following the course of isoniazid therapy.

Scenario 2: active pulmonary TB in the mother
Mother has active pulmonary TB and is infectious at time of delivery.

- Infection control procedures are required to reduce risk of transmission to baby, family, visitors and health staff. This would normally include admission to a well-ventilated isolation unit, education on cough hygiene and wearing of respiratory isolation face-mask while still infectious when in contact with others, e.g while handling or breastfeeding baby
- There is no need to separate mother and infant if the infant is taking isoniazid and infection control measures are being used. There is no contraindication to infants rooming-in with their mother in hospital. Only separate mother and infant in highly exceptional circumstances, e.g. if there is a high suspicion of MDR-TB or if the mother is unable to care for her child.
- Assess neonate for clinical, laboratory or radiological evidence of congenital TB and treat with multiple drug therapy if it is present.
- In the absence of active disease give isoniazid (15 mg/kg/day) to the newborn.
- It is advised to withhold BCG immunisation. This is not because of evidence of reduced effectiveness but because it may interfere with clinical diagnosis of TB disease in the baby.
- After 6 weeks of isoniazid, do a TST and repeat CXR.
  - If TST non-reactive (< 5 mm) and CXR normal, continue isoniazid and repeat these at 12 weeks and 6 months.
  - If TST is reactive (≥ 5 mm) at 6 weeks, 12 weeks or 6 months of age, investigate thoroughly for pulmonary and extrapulmonary disease. If evidence of disease is present, treat with multiple drug therapy. If there is no evidence of pulmonary and extrapulmonary disease, continue isoniazid to complete a 9-month course.
- After 6 months of isoniazid, do a TST and repeat CXR.
  - If TST is non-reactive and the CXR is normal, discontinue isoniazid if the mother is now sputum smear-negative, and give BCG to the infant. The reason for repeating TST at 6 months and giving isoniazid up to

this time even if TST was non-reactive at birth is that tuberculin conversion may be delayed for up to 6 months after infection at birth.
  - If TST is non-reactive at 6 months of age but the mother is still sputum smear-positive, do not stop isoniazid. The mother needs investigation for treatment failure and possible drug resistance. The baby should be continued on isoniazid unless isoniazid resistance is thought to be likely or confirmed; if so, rifampicin would have to be given. A strong case exists for BCG in this situation.
- If BCG is given to the baby (an option that some advocate at birth), watch for accelerated vaccine response.
  - An accelerated response suggests that the baby has been infected. Follow the action outlined above; consider an accelerated response as you would a positive TST.
- If there is no accelerated response, a tuberculin skin test can be done 2–4 weeks later. A tuberculin reaction at this early stage suggests that it is due to a natural infection and not to the BCG.

Scenario 3: non-infectious pulmonary TB in the mother

Mother is on treatment for pulmonary TB but is no longer infectious (sputum culture is negative) at the time of delivery.

- Assess neonate for clinical, laboratory or radiological evidence of congenital TB (see above), and treat with multiple drug therapy if it is present.
- No need for separation, but infection control measures such as wearing of respiratory isolation face-mask would be advised until mother has completed at least 8 weeks of TB treatment.
- Examine the infant at monthly intervals.
- Evaluate TB risk in family members.
- Do a TST at 6 weeks, 12 weeks and 6 months (see above for further action).
- There is a case for isoniazid preventive therapy for 6 months even if the mother is now sputum-negative because:
  - the mother might have had haematogenous or genital spread resulting in infection of the infant
  - the mother may still be infectious.

Scenario 4: mother's treatment completed successfully

Mother completed antituberculous treatment during pregnancy and is no longer infectious.

- No separation of mother and baby is required.
- Evaluate TB risk in family members.
- Give BCG to the newborn, unless future risk of TB exposure is considered low.

Scenario 5: TB in the family
Another member of the family is being treated for TB.

- If the family member with TB has completed treatment:
  - evaluate the family member before the baby returns home
  - give BCG to the newborn.
- If the family member is on treatment:
  - no contact with patient for at least 8 weeks after being culture negative
  - give BCG to the newborn.
- If the family member is infectious:
  - the best course of action is to have no contact with the infectious person for at least 8 weeks after being culture negative
  - if exposure is unavoidable or likely, give isoniazid until the index case is culture-negative for 8 weeks
  - an alternative is to give BCG to the newborn before return to household.

Scenario 6: hospital exposure of baby to TB
Baby has been exposed to a healthcare worker with infectious TB while in the nursery.

- Infection is rare under nursery conditions, but it can and does occur. Important information will include closeness and duration of contact, smear and culture result of index case and drug susceptibility pattern.
- Investigate for TB, i.e. clinical features, TST and CXR. If no evidence of active TB disease is found, give isoniazid to the newborn for 6 months.
- Follow up monthly.
- If there is clinical suspicion of TB at any stage, then investigate for TB disease and treat as needed.
- If the baby is well at 6 months, TST negative and CXR normal, discontinue isoniazid and review at 12 months.
- If the baby is well at 6 months, but TST reactive (≥ 5 mm), continue isoniazid until 9 months and review at 12 months.

4.6 Screening for tuberculosis during pregnancy

Routine screening for tuberculosis during pregnancy is not necessary. However, screening should be considered for several groups of pregnant women.

4.6.1 Women at high risk

Women at high risk are those with possible active TB, or at high risk of latent TB and disease progression. These include:

- Patients with symptoms suggestive of TB. Investigate for active disease; do not delay CXR.
- HIV-infected patients (and other profoundly immunocompromised patients) – TST/QFN-GIT and CXR are indicated if not done previously or if the woman has had possible exposure to TB since the last test.
- Recent close contact of infectious TB. Perform TST/QFT-GIT, followed by a CXR if this is positive.

4.6.2 Women at moderate risk

Women with moderate risk of disease progression include people with epidemiology suggestive of previous but possibly remote TB exposure, such as recent arrival from a country with a high TB incidence. For these women, perform TST/QFN-GIT, clinical assessment and CXR after the twelfth week of pregnancy.

4.7 Latent TB infection (LTBI) in pregnancy

For those with latent TB infection, including in HIV-infected women, there is no evidence that pregnancy increases the chance of TB reactivation. The TST reaction is not altered by pregnancy. The effect of pregnancy on QFN-GIT has not been evaluated; however it is recommended for use in all current circumstances in which TST is currently used.

There is no fetal toxicity from isoniazid (or rifampicin); however, the risk of isoniazid hepatitis is higher in women compared with men, and is possibly higher in the post-partum period. Treatment of LTBI is usually withheld until 3 months after pregnancy unless the patient has been recently infected (within the last 2 years), is HIV-infected, or has medical conditions that increase the risk for reactivation of inactive TB.

If women are to be treated during pregnancy, careful screening, patient education about side-effects and monthly questioning for symptoms of toxicity (especially hepatitis) are recommended. Pyridoxine should be prescribed routinely, at a dose of 50 mg daily. Monthly hepatic enzyme testing should be performed if isoniazid therapy is given in pregnancy.

4.8 Infertility related to tuberculosis

Genital tuberculosis is a well-known contributing factor to female infertility. While in-vitro fertilisation techniques can be successful in women with genital TB, it appears that outcomes are less favourable compared with other aetiologies of tubal infertility. A recent study evaluating the impact of LTBI in patients with unexplained infertility and previous unsuccessful IVF attempts suggests that improved outcomes may be obtained in women with latent infection who undergo treatment with antituberculous therapy.

References and further reading

Bass JB. Tuberculosis in pregnancy. In: UpToDate, Basow DS (Ed) UpToDate, Waltham MA, 2012.

Bothamley G. Drug treatment for tuberculosis during pregnancy: safety considerations. Drug Saf 2001; 24:553–65.

Dam P, Shirazee HH, Goswami SK, et al. Role of latent genital tuberculosis in repeated IVF failure in the Indian clinical setting. Gynecol Obstet Invest 2006; 61:223–7.

Doveren RFC, Block R. Tuberculosis and pregnancy – a provincial study (1990–1996). Netherl J Med 1998; 52:100–6.

Efferen LS. Tuberculosis and pregnancy. Curr Opin Pulm Med 2007; 13:205–11.

Espinal MA, Reingold AL, Lavandera M. Effect of pregnancy on the risk of developing active tuberculosis. J Infect Dis 1996; 173:488–91.

Franks AL, Binkin NJ, Snider DE, Jr., et al. Isoniazid hepatitis among pregnant and postpartum Hispanic patients. Public Health Rep 1989; 104:151–5.

Jana N, Vasishta K, Saha SC, Ghosh K. Obstetrical outcomes among women with extrapulmonary tuberculosis. N Engl J Med 1999; 341:645–9.

Chapter 5

Diagnosis and management of latent TB infection

5.1	Decision-making in testing and treatment of latent TB infection	63
5.2	Pre-test probability of TB infection	64
5.3	Risk of TB disease progression following TB infection	64
5.4	Interpretation of LTBI test results	65
5.5	Treatment efficacy and adherence to therapy	71
5.6	Risk of adverse effects from LTBI treatment	72
5.7	Practical steps in management of LTBI	74
5.8	Key concepts	78

5.1 Decision-making in testing and treatment of latent TB infection

When deciding whether to treat for latent TB infection (LTBI), the first consideration is to rule out active tuberculosis by taking a careful history, examining the patient and reviewing a recent CXR. When active TB is excluded, the following questions must be considered.

- How likely is it that the patient has LTBI, based on epidemiological grounds (pre-test probability)?
- What is the probability that the patient will go on to develop active tuberculosis?
- What are the test results and their interpretation?
- What is the efficacy of treatment in this patient, taking into account likelihood of treatment adherence and completion?
- What is the propensity for the patient to develop an adverse reaction (largely age-dependent hepatotoxicity) to treatment for LTBI?

Table 5.3.1 Risk of TB disease in 2 years following TB exposure and infection

Age group	Risk (%)
Infants ($\leq$ 1 year)	50
Children (1–2 years)	12–25
Children (2 to 5 years)	5
Children (5–10 years)	2
Adolescents and young adults	10–20
Older adults	3–5

5.2 Pre-test probability of TB infection

Pre-test probability of TB infection must be assessed. Groups with a high probability of having TB infection include:

- immigrants and refugees from countries with a high prevalence of TB – this is the most common risk factor for LTBI in Australia
- those with a high degree of exposure – close contacts of smear-positive pulmonary TB cases (family members, household contacts and other contacts with more than 8 hours of exposure), healthcare workers with exposure to TB patients in high-prevalence countries
- other specific high-risk groups – these may include the homeless, some Indigenous communities, some healthcare workers, travellers and expatriates who have spent prolonged periods of time in TB-endemic countries, and on occasions those in prisons and institutions.

5.3 Risk of TB disease progression following TB infection

Once a person has contracted LTBI, the risk for progression to TB disease varies. The greatest risk for progression to disease occurs within the first 2 years after infection, when approximately half of the 5–10% lifetime risk occurs. However this risk is age-dependent (see Table 5.3.1).

Numerous clinical conditions also are associated with increased risk for progression from LTBI to TB disease. These conditions can be divided into those with a more than 5-fold risk, and those with a 1–5-fold risk, relative to the risk in people who have acquired latent TB infection more than 2 years previously with no other risk factors. HIV infection is the strongest known risk factor.

5.3.1 More than five times relative risk
- People with HIV infection
- Infants and children aged < 5 years
- People receiving tumour necrosis factor-alpha (TNF-α) inhibitors, e.g. infliximab, adalimumab
- People who were recently infected with *M. tuberculosis* (within the past 2 years)
- People with a history of untreated or inadequately treated active tuberculosis, including people with fibrotic changes on chest radiograph consistent with prior active tuberculosis (not just a granuloma)
- People with medical conditions such as silicosis, chronic renal failure, leukaemia, lymphoma, or cancer of the head, neck, or lung
- People who have had a jejunoileal bypass

5.3.2 One–five times relative risk
- People who have had a gastrectomy
- People who weigh < 90% of their ideal body weight
- Cigarette smokers and people who abuse drugs or alcohol
- People with diabetes
- People receiving immunosuppressive therapy such as > 15 mg prednisolone daily for more than 1–2 months, or etanercept (associated with a lower relative risk compared with the TNF-α monoclonal antibodies)
- Recently arrived refugees from countries with a high TB prevalence.

If active infection does occur, those with HIV infection and children < 5 years old also have a greater risk of poor outcome, including meningitis and disseminated disease and death.

5.4 Interpretation of LTBI test results

The two most widely used tests for diagnosis of LTBI in Australia are the tuberculin skin test (TST) and the Quantiferon Gold-TB In-Tube (QFN-GIT) test, an interferon gamma release assay (IGRA). Neither test has perfect sensitivity or specificity, but the QFN-GIT has many advantages over TST that are elaborated below. In VIDS, QFN-GIT is the preferred test, but the TST is still widely used in other settings. It is not uncommon to see a patient who has had both tests, with results that are discordant (20%). The following subsections describe TST and

QFN-GIT, and their advantages and disadvantages and interpretation, including how to interpret discordant results.

5.4.1 Tuberculin skin test (TST)
Disadvantages of TST
Poor specificity: false positive TST results
TST is known to have high rates of false positives. Factors that lead to a false positive TST include:

- Previous BCG vaccination. A single BCG vaccination at birth usually leads to a positive TST that wanes over the next decade. However, BCGs given after age 2 and repeat BCGs can lead to prolonged (false) positive TST.
- Exposure to non-tuberculous mycobacteria (NTM). Geographically clustered in tropical and sub-tropical regions, exposure to these environmental mycobacteria can lead to a false positive TST.

Poor sensitivity: false negative TST results
Immune dysfunction and other factors cause false negative TST results. Examples include:

- HIV, and other immunosuppressive disease or therapy
- inadequate nutrition
- malignancy
- active TB, especially severe/miliary TB
- concurrent viral infection
- children and elderly

Tuberculin reversion
Reversion from a previous positive to a negative reaction occurs in 2–20% of people over 2 to 20 years and is more likely in those with an initial reaction of 10–14 mm, i.e. not a strongly positive reaction. With continuing TB exposure a large tuberculin reaction tends to be maintained.

Individual and interpreter variability
The TST response in an individual is highly variable. Two tests done at the same sitting by the same operator on different arms may show 15% discordance. The same tuberculin reaction in an individual measured by two different experienced operators may also show 15% discordance in the readings. Under-reading of a positive TST was common in one study: 33% failed to identify a positive reaction.

Booster reactions
In patients with previous mycobacterial infection (either BCG vaccination, TB or NTM), an initial TST may be negative, but if this is followed by a second TST, the protein from the first TST boosts the immune response and can lead to a positive second test if performed from 1 week to 1 year (and possibly longer) later.

The implication of this is that when people have multiple TSTs (e.g. serial testing in high-risk healthcare workers, or screening before and after travel to a country with a high TB incidence), a positive result on a second or subsequent TST following an earlier negative result could be erroneously interpreted as indicating TB exposure and infection in the interval between the tests, whereas the result is really due to boosting of a waned immune response to a previous mycobacterial infection. This issue is not relevant if serial testing is performed with IGRA.

Interpretation of TST without IGRA
The current interpretation of TST results uses several cut-offs, with higher cut-off for those with low pre-test probability (≥ 15 mm) and lower cut-off (≥ 5 mm) for those with high probability of progression to tuberculosis. The intermediate cut-off (≥ 10 mm) is for those with epidemiological risk factors suggesting high pre-test probability and with those who have greater than average probability of progression to tuberculosis who do not fall into the high-risk group above.

Note that the choice of cut-offs is based on factors other than TST performance in these groups. This makes the interpretation very confusing, especially in the setting of discordant QFN-GIT results as discussed below.

5.4.2 Interferon gamma release assays (IGRA)
Interferon (IFN)-gamma release assays (IGRAs) are whole blood ex-vivo T cell-based assays for the detection of LTBI. The principle of the assay is that T cells of individuals previously infected with *M. tuberculosis* will produce interferon-gamma when they encounter TB-specific mycobacterial antigens. Thus, a high level of interferon-gamma production is presumed to be indicative of TB infection. The antigens used in IGRAs (ESAT-6, CFP-10 with or without TB 7. 7) are encoded by *M. tuberculosis* genes that are not shared with *M. bovis*-BCG or most NTM, and hence they are called 'region of difference' (RD) antigens.

There are two methods for detecting the IFN-gamma released by the T cell: an enzyme-linked immunosorbent assay (ELISA, e.g. QFN-GIT), and an enzyme-linked immunospot assay (ELISPOT, e.g. T-SPOT.*TB*). The QFN-GIT is the more widely used of these tests in Australia. In both tests, a control mitogen and a nil sample are tested in parallel with the *M. tuberculosis* antigens. IGRA tests are reported as positive, negative or indeterminate. Indeterminate tests may represent low response to the control antigen or high response to the nil control. Currently there is a single cut-off at which the test is regarded as positive.

Advantages over TST
- Better specificity in the setting of BCG vaccination and NTM exposure (most common in tropical and subtropical regions).
- Similar sensitivity to TST in most cases, although IGRAs may be more sensitive in the setting of immune suppression, e.g. in haematology and HIV-infected patients.
- Simple, single blood test, not requiring return visit (hence definitely preferred in groups in whom follow-up TST reading may be difficult).
- Automated interpretation of QFN-GIT eliminates problems with inter-user reliability.
- No booster phenomenon, as the individual does not encounter antigen in this *in vitro* test. (Results of studies of whether IGRA response can be boosted by TST are inconsistent.)

Disadvantages
- IGRAs are more expensive than TST, so their use is not feasible in many of the countries in which TB incidence is highest.
- The accuracy of IGRAs appears to be at least as good as TST in almost all settings. However, IGRAs may not perform as well in children as in adults, and currently the CDC suggest that in children < 5 years TST is preferred, although QFN-GIT is an acceptable alternative. Current practice at the Royal Children's Hospital in Melbourne (February 2012) is still to perform TST for those under 16 years.
- IGRAs require a phlebotomy and correct specimen handling and processing.
- There is relatively limited data concerning the ability of a positive or negative IGRA test to predict the subsequent development or absence of TB (although such data are accumulating and indicate at least equivalent predictive performance to the TST).

Interpretation of QFN-GIT
- A single cut-off value (TB antigen minus nil control ≥ 0.35 IU/mL) is used to define a positive result. Some experts have suggested applying different

cut-off values for high- and low-risk populations (as is done with TST), or defining a 'borderline' range, but this is not current practice.
- An indeterminate result can be due to a low positive control or a high negative control (see section 5.4.4).

Factors affecting sensitivity
- As with TST, IGRAs are less sensitive in the setting of significant immune suppression. However, studies in HIV infection indicate they are less likely to be falsely negative than TST.

Factors affecting specificity
- False positives may occur in the presence of three NTM: *M. marinum, M. szulgai, M. kansasii*.

5.4.3 Interpretation of discordant TST and IGRA
Without a gold standard it is very hard to determine sensitivity and specificity, but the following considerations help guide interpretation.
- In immune-competent adults, sensitivity of QFN-GIT and TST are both estimated to be around 70–80%. TST specificity is high (over 90% if 15 mm used as cut-off) in those not vaccinated with BCG or exposed to NTM. QFN-GIT is 93–99% specific in immunocompetent people whether or not they are BCG vaccinated. In patients with a low risk of exposure, discordant TST+/IGRA− results are common, especially in those previously vaccinated with BCG.
- In HIV-positive patients, TST is more likely than the T-SPOT.*TB* assay to produce a false negative result, and the lower the CD4 cell count the more likely this is to occur. QFN-GIT seems to have lower sensitivity than T-SPOT.*TB* but is more sensitive than TST. QFN-GIT has higher rates of indeterminate tests due to negative mitogen control response when the CD4 cell count is ≤ 100 per microlitre compared with when CD4 cell count is >100 per microlitre.
- In the elderly TST is more likely to be falsely negative, whereas QFN-GIT is less affected by age.
- Comparative studies are less common in children. QFN-GIT has been found to be less sensitive than T-SPOT *TB* in children with active TB.
- Following treatment of active or latent TB, reversion from a positive to a negative result can occur with both TST and IGRAs. The significance of

this is unknown, in particular whether it indicates eradication of all *M. tuberculosis* organisms.

Meta-analyses have shown that IGRA specificity is consistently high in those with and without prior BCG vaccination, whereas TST specificity is diminished by prior BCG vaccination. A fully integrated statistical meta-analysis, which included latent class and mixed effects, estimated that sensitivity for IGRAs was 64% and specificity was 99.7%. TST specificity was similar to IGRA in those without prior BCG but fell to 50% in those with prior BCG.

IGRA positive, TST negative
It is most likely that this discordance represents a falsely negative TST because data show that for groups such as the elderly and the immunocompromised IGRAs are more sensitive than TST.

IGRA negative, TST positive
This pattern of discordance is relatively common and is usually due to a falsely positive TST. If the patient has risk factors for a false positive TST, such as recent BCG, multiple BCGs, BCG at older age or exposure to NTM, the probability that this is a false positive TST is high.

If the TST reaction is large (> 20 mm) however, it is rarely a false positive. In such cases, the IGRA may be a false negative, particularly if the TB mitogen result is close to the positive cut-off.

5.4.4 Management of indeterminate QFN-GIT result
Indeterminate QFN-GITs should be repeated (once).

An indeterminate QFN-GIT due to a high nil control value (reflecting elevated background T-cell activity) may well lead to a more definitive result in the second test. An indeterminate result due to reduced mitogen activity is more likely to remain indeterminate. After two indeterminate results, judge the situation by epidemiology and risk of progression. A TST can be done in cases where: (i) the patient has no risk factors for false positive TST (no BCG and/or low likelihood of exposure to NTM); (ii) the TST is expected to be positive if LTBI exists (not significantly immunosuppressed); and (iii) a positive TST would lead to a decision to treat the patient. Examples include recent significant TB exposure in an Australian-born person, or someone about to undertake TNF-α inhibition therapy.

5.4.5 IGRA testing in those who have had a TST
If a patient has already had a TST, an IGRA should only be done if interpretation of the TST result is unclear and the result of the IGRA would affect management.

The following guidelines will assist, but do not cover every conceivable situation. In practice, epidemiology (pre-test probability), risk of progression to disease and risk factors for false positives (especially prior BCG) and negatives all have to be taken into account.

- If the TST is ≥ 15 mm, IGRA not indicated.
- If the TST is ≥ 10 mm and < 15 mm:
  (a) if the person has a history of TB exposure *or* has suggestive X-ray changes (e.g. > 1 calcified nodules, upper lobe fibrosis) or is immunosuppressed: regard TST as positive and IGRA not indicated. This recommendation is stronger if person has *not* had a BCG.
  (b) if the person has none of risk factors in (a): regard TST as indeterminate and IGRA indicated. This recommendation is stronger if person *has* had a BCG.
- If the TST is < 10 mm, IGRA not indicated, unless the person has risk factors for LTBI *and* progression to active TB (e.g. patient from TB-endemic country with HIV, or patient with abnormal CXR about to start infliximab).

5.5 Treatment efficacy and adherence to therapy

Patients diagnosed with latent TB according to the recommendations discussed above should be considered for treatment with isoniazid monotherapy. Before recommending treatment in older patients and in those with complicating factors, consideration should be given to the probability of disease progression to active TB if untreated, likely compliance and treatment efficacy, and the risk of adverse drug reactions.

5.5.1 Treatment efficacy

Isoniazid is the recommended first-line treatment for LTBI at a dose of 300 mg once daily. The aim should be to treat for 9 months but in some cases therapy may be ceased at 6 months, for example if the patient is not motivated to continue therapy or is experiencing minor drug side-effects. HIV-positive and other immunosuppressed patients and those with fibrotic lesions on CXR should always be treated for for 9 months.

Studies have shown that both a 6-month and a 12-month course are effective at reducing progression of disease (RR 0.44 and 0.38 respectively), with no statistically significant difference between the progression rates to active TB in the two groups. However, in subgroup analyses of several trials the maximal beneficial effect of isoniazid is likely to be achieved by 9 months, and minimal additional

Table 5.5.1 Evidence for LTBI therapy

Drugs	Duration (months)	Interval	HIV-negative	HIV-positive
Isoniazid	9	Daily	A(II)	A(II)
	6	Daily	B(I)	C(I)
Rifampicin	4	Daily	B(II)	B(III)

A = preferred; B = acceptable alternative; C = offer when A and B cannot be given. I = randomised clinical trial data; II = data from clinical trials that are not randomised or were conducted in other populations; III = expert opinion.

benefit is gained by extending therapy to 12 months. When compared with placebo, both 6-month and 12-month regimens are effective in HIV-positive patients, with no direct comparison available (see Table 5.5.1).

Rifampicin for 4 months is an acceptable alternative if there is significant toxicity to isoniazid or if the index case has known isoniazid-resistant TB. There are no head-to-head long-term efficacy studies comparing 4R with 9H.

Isoniazid and rifampicin given in combination for 3 months is another effective regimen; it is used in the United Kingdom but not commonly in Australia.

Rifampicin and pyrazinamide is effective and only has to be given for 2 months, but is not recommended because of high rates of hepatotoxicity.

Just as this handbook was in the final stages of its preparation, a regimen of directly observed **weekly isoniazid and rifapentine** (the latter is a rifamycin agent with a long half-life) for 3 months was reported to be as effective as a 9-month course of isoniazid. Rifapentine is currently not licensed for use in Australia.

5.5.2 Adherence

It is difficult to anticipate which patients will be compliant with therapy, but this is an important consideration in the final decision about treatment and treatment duration. Patients should ideally express a wish to take the medication and be willing to commit at the outset to daily medications for the duration of the prescribed treatment, and to attend the outpatient clinic regularly for monitoring.

5.6 Risk of adverse effects from LTBI treatment
5.6.1 Hepatitis
- Abnormal LFTs are common in patients following commencement of isoniazid therapy. Elevations of AST or ALT occur in 10% to 20% of people on isoniazid; however, clinical hepatitis is infrequent and is directly

related to increasing age – a very low risk below the age of 35, but a significant risk in those over the age of 50, according to a study conducted in the 1970s. A more recent study suggests the rate of symptomatic hepatotoxicity may be one-fifth as common as previously believed in those over 35 (see Table 5.6.1).
- Another recent study of people over 35 years showed that of 335 patients commencing isoniazid for LTBI, 12% developed AST/ALT > 2× the upper limit of normal, but none developed symptomatic hepatitis and only one required temporary cessation of therapy.
- The risk of hepatitis is increased in those with pre-existing liver disease and increased alcohol intake. Asymptomatic chronic carriers of hepatitis B (HBsAg+) do not have an increased risk of isoniazid hepatitis compared to other people of similar age, although caution is needed in patients with chronic HBV who develop abnormal LFTs. Several studies have also suggested that being of white race and female trended towards increased risk, but these have not been statistically significant.
- Isoniazid hepatitis can be fatal: the mortality rate is 6–8% and increases with age and alcohol intake.
- While hepatitis is more common in the first 2 months of therapy, approximately 50% of elevated LFTs occur after this date. Seventeen severe isoniazid-associated liver injury events were reported by the CDC (years 2004–08), of which 16 occurred after the first month of therapy.

5.6.2 Peripheral neuropathy
Peripheral neuropathy occurs in < 1% of people on isoniazid and is preventable with pyridoxine (vitamin B6) supplements of 25 mg/day. Pyridoxine is indicated for people older than 65 years of age, pregnant women, people with diabetes, daily alcohol users, people with chronic renal failure, poorly nourished people (frequent in recently arrived immigrants and refugees) and anyone with any other predisposition to peripheral neuropathy; otherwise we do not prescribe pyridoxine routinely.

5.6.3 Other isoniazid adverse effects
- Allergic rash
- Neuropsychological effects such as minor difficulties with concentration and dizziness – usually avoided by administering the dose at night
- Acne and minor alopecia
- Gastrointestinal upset

Table 5.6.1 Risk of hepatitis in patients treated with isoniazid by age, according to two different studies

Age (years)	Hepatitis risk (%) (from 1970s study)*	Age (years)	Hepatitis risk (%) (from 1990s study)**
< 20	Rare	< 15	< 0
20–34	up to 0.3	15–34	0.8
35–49	up to 1.2	35–64	0.21
50–64	up to 2.3		
≥ 65	4.6	> 65	0.28

* Kopanoff et al (1978)
** Nolan et al (1999)

- Drug interactions – increase in serum levels of phenytoin and disulfuram

5.7 Practical steps in management of LTBI
5.7.1 Who to screen for LTBI
Testing for LTBI should be targeted at those with risk factors for TB infection or progression to active TB. Groups for whom testing is recommended are:

- Contact tracing of active TB cases.

    This is undertaken by the public health TB program. In Victoria the DoH TB Control Program uses TST. A baseline TST is done at the time of contact tracing when a significant contact is established (> 8 hours contact in smear-positive pulmonary TB, and 24–48 hours if smear-negative pulmonary TB). This is followed by a repeat test in 8–12 weeks after last infectious contact if the initial test is negative. An argument can also be made for just doing one test at 8 weeks, particularly in those with no prior exposure to TB.

- Refugee screening.

    Testing is recommended for all 12–35-year-old refugees except those with a known history of active tuberculosis, ideally within 12 months of arrival but later if not done earlier. Refugees have a high risk of recent infection or re-infection. Therefore, depending on the epidemiology, the age cut-off for routine testing may be higher in this group.

- Immigrant and onshore visa applicant screening.

    In Victoria, limited screening of selected recently arrived immigrants and visa applicants applying from within Australia – usually those with

changes suggestive of inactive TB on the health check CXR – is done by the Migrant Screening Clinic, using TST.
- Healthcare worker and student screening.

  Guidelines recommend pre-employment screening for all employees with patient contact, and annual screening for employees at high risk of ongoing TB exposure. Healthcare students should also be screened prior to beginning clinical placements. Those born in a TB-endemic country are a particular priority for this testing.
- Patients who are to undergo immunosuppressive treatment.

  Testing for latent TB infection is recommended for patients treated with TNF-α inhibitors, high doses of corticosteroids and high-dose chemotherapy, and for organ transplant patients.
- Patients with HIV infection.
- Patients with silicosis.
- Other groups.

  Testing for latent TB may be considered in other groups at higher risk of TB infection or progression to active disease, such as the homeless and prisoners.

At VIDS, patients who require screening for latent TB undergo testing with QFN-GIT.

5.7.2 Exclude active infection

Active infection is treated with multiple drugs and must first be excluded.

- Does the patient have any symptoms or signs? These may include constitutional or respiratory symptoms, back pain (spinal disease) and enlarged lymph nodes.
- Are there radiological changes? Are there old CXRs for comparison? If so, have these radiological changes been present for > 2 years? A CXR showing no abnormalities consistent with active tuberculosis is mandatory prior to initiating treatment. Stable chest films for at least 1 year would also be evidence for inactive disease, but, if in doubt, sputum should be obtained. One or two calcified granulomas do not usually qualify as a significant TB abnormality. People with an isolated granuloma have a similar risk of developing active disease as those with completely normal films.
- Collect sputum or bronchial specimens for culture in those with suspected pulmonary tuberculosis.

- Even if all smears are negative and the patient is asymptomatic, the possibility of asymptomatic TB disease remains. Consider bronchoscopy if radiographic changes are not typical for old scarred TB and cannot be proven to be stable, or if the patient has respiratory symptoms even in the absence of worrying CXR changes. Consider treatment for TB disease in the absence of X-ray changes suggestive of active disease if (i) the patient has constitutional symptoms such as weight loss or fever, or (ii) the patient has silicosis.

5.7.3 Determine if treatment of latent TB infection is indicated

Recommend treatment, regardless of age, for those patients at highest risk of progression to active TB:

- Probable or confirmed recent infection following recent TB exposure
- Conversion from negative to positive on serial LTBI testing
- HIV infection
- Significant immunosuppressive therapy – organ transplantation, TNF-α inhibitor therapy, high-dose chemotherapy
- Silicosis
- Children.

Consider treatment for patients at lower but still increased risk of progression to active TB:

- Significant CXR changes indicative of past TB infection
- Refugees < 45 years
- Lesser degrees of immunosuppression
- Some medical conditions such as chronic renal failure.

The decision to treat or not will depend on balancing the risk of TB progression against the risks of treatment (chiefly hepatotoxicity) and likelihood of adherence.

Treatment is generally not indicated for patients older than 35 years with no risk factors for TB progression. One exception is a healthcare worker, because of the potential for patient and staff exposure should that person develop TB.

Some people with LTBI are followed with serial CXRs, although this is an insensitive way of detecting early TB reactivation. This approach can be taken if a patient declines a recommendation for LTBI treatment, if the risks of treatment outweigh the risk of reactivation of latent TB, or if a patient cannot be monitored adequately for adverse drug reactions and adherence.

If serial CXRs are conducted, they are usually done 0, 6 months, 12 months and 24 months after a person is believed to have been exposed to TB, and are therefore not required for those with LTBI likely to have been acquired >2 years prior to review.

5.7.4 Determine medical management
- Isoniazid monotherapy is the preferred agent, with rifampicin monotherapy an acceptable alternative in selected patients as discussed above.
- Special groups:
    - Children: as for adults, treat for 9 months with isoniazid.
    - Pregnant women: usually defer treatment until after delivery except when the risk of TB developing during pregnancy is high; e.g. the patient is HIV-positive or has had a very recent exposure (within months). There is no risk of isoniazid to the fetus, but there may be a risk of flare of hepatitis in the mother following delivery.
    - Suspected MDR-TB latent infection. Currently we do not change management based on concerns about rates of MDR-TB in the patient's country of origin. However, this is under review as the epidemiology changes. If the case contact were known to have MDR-TB, the principle of management is to treat with at least two drugs to which the TB is known to be susceptible. Based on current evidence including local mycobacterial susceptibility data, we suggest moxifloxacin plus either ethambutol or pyrazinamide if empiric therapy is to be commenced while extended sensitivities are awaited. Optimal treatment duration is not established, but our current practice is for 6–9 months of therapy.

5.7.5 Baseline testing before starting LTBI treatment
- Prior to initiating LTBI treatment, check LFTs, HBsAg, HCV Ab and serum vitamin D level. It is important to document baseline LFTs to assess any abnormalities that may develop on treatment. Chronic hepatitis B and C affect the later interpretation of LFTs. Vitamin D deficiency is extremely common in TB clinic patients.
- Provide patient with information sheets (in Victoria, these are produced by the TB Control Program and are available in several languages).

5.7.6 Monitoring during LTBI treatment
- Patients should first be reviewed no later than 4 weeks after starting treatment for LTBI. Subsequent appointments can be after 2 then 3 months so

the patient has an appointment at months 0, 1, 3, 6, and 9. (Note that this is less frequent than the monthly follow-up recommended in US guidelines.) The final appointment is useful to ensure and document full compliance with therapy and to remind the patient of the small ongoing risk of active TB.
- Those considered at high risk of adverse events or treatment non-completion should have monthly appointments and LFTs. These are patients:
  - aged over 35 years
  - with a history of daily alcohol intake, abnormal baseline liver function tests, known chronic liver disease or receipt of hepatotoxic medications
  - where there are concerns regarding adherence
  - with HIV infection and pregnant/post-partum patients.
- At each outpatient clinic visit:
  - Assess adherence.
  - Evaluate for signs and symptoms of active TB and drug reactions.
  - Remind patients of signs and symptoms of hepatotoxicity.
  - Repeat liver function tests for patients listed above or for those with GI symptoms.
- Routine CXR, TST and IGRA are neither necessary nor useful during or after treatment of latent infection.
- If a patient misses one or more doses of medication:
  - Interruption of < 1 month is probably not significant. Ensure the complete course is taken.
  - Interruption for > 1 month in the first 3 months of treatment: start the whole course again and ensure future compliance.
  - Interruption for > 1 month after 3 months: ensure future compliance and complete the planned treatment course.

5.8 Key concepts
- Testing for LTBI should be targeted at those with risk factors for TB exposure or progression to active TB.
- The TST is limited by its poor specificity, especially in BCG-vaccinated people, and by practical considerations relating to test administration and reading of the result.

- IGRAs have improved the specificity of testing for LTBI and offer considerable practical advantages over the TST. In adults, IGRAs can be used in all situations in which TST is currently used, and IGRAs are preferred in BCG-vaccinated people.
- Neither the TST nor the new IGRAs should be used for diagnosis of active TB disease.
- Treatment of LTBI is targeted at those with a high risk of progression to active TB: close contacts of smear-positive pulmonary TB, recent TST or IGRA converters, young children, HIV-infected and other significantly immunosuppressed patients, recently arrived refugees from TB-endemic countries, and patients with significant CXR changes or silicosis.
- Treatment of LTBI is with isoniazid for 9 months. For selected patients, an alternative is rifampicin for 4 months.
- Combination therapy with isoniazid and rifapentine given weekly for 3 months is a very promising treatment approach (but rifapentine is currently unavailable in Australia).

References and further reading

Centers for Disease Control and Prevention. Severe isoniazid-associated liver injuries among persons being treated for latent tuberculosis infection – United States, 2004–2008. MMWR 2010; 59(8):224–9.

Centers for Disease Control and Prevention. Targeted tuberculin testing and treatment of latent tuberculosis infection. MMWR 2000; 49(RR-6):1–51.

Centers for Disease Control and Prevention. Updated guidelines for using interferon gamma release assays to detect *Mycobacterium tuberculosis* infection – United States, 2010 MMWR 2010; 59(RR-5):1–28.

Comstock GW. How much isoniazid is needed for prevention of tuberculosis among immunocompetent adults? Int J Tuberc Lung Dis 1999; 3:847–50.

Diel R, Loddenkemper R, Meywald-Walter K, et al. Comparative performance of tuberculin skin test, QuantiFERON-TB-Gold In Tube assay, and T-Spot.TB test in contact investigations for tuberculosis. Chest 2009; 135:1010–18.

Ferebee SH. Controlled chemoprophylaxis trials in tuberculosis. A general review. Bibl Tuberc 1970; 26:28–106.

Gilroy SA, Rogers MA, Blair DC. Treatment of latent tuberculosis infection in patients aged > or =35 years. Clin Infect Dis 2000; 31(3):826–9.

Horsburgh CR, Jr, O'Donnell M, Chamblee S, et al. Revisiting rates of reactivation tuberculosis: a population-based approach. Am J Respir Crit Care Med 2010; 182(3):420–5.

Kopanoff DE, Snider DE Jr, Caras GJ. Isoniazid-related hepatitis: a U.S. Public Health Service cooperative surveillance study. Am Rev Respir Dis 1978; 117:991–1001.

Nolan CM, Goldberg SV, Buskin SE. Hepatotoxicity associated with isoniazid preventive therapy: a 7-year survey from a public health tuberculosis clinic. JAMA 1999; 281:1014–18.

Pai M, Zwerling A, Menzies D. Systematic review: T-cell-based assays for the diagnosis of latent tuberculosis infection: an update. Ann Intern Med 2008; 149(3):177–84.

Sterling TR, Villarino ME, Borisov AS, et al. Three months of rifapentine and isoniazid for latent tuberculosis infection. New Engl J Med 2011; 365:2155–66.

Vinton P, Mihrshahi S, Johnson P, et al. Comparison of QuantiFERON-TB Gold In-Tube Test and tuberculin skin test for identification of latent *Mycobacterium tuberculosis* infection in healthcare staff and association between positive test results and known risk factors for infection. Infect Cont Hosp Epi 2009; 30(3):215–21.

Chapter 6

HIV and tuberculosis

6.1	Testing for HIV infection in TB patients	81
6.2	Screening for latent TB infection in HIV patients	82
6.3	Treating latent TB in HIV patients	82
6.4	Diagnosing active TB in HIV patients	83
6.5	Treating active TB in HIV patients	84
6.6	Antiretroviral regimens in patients being treated for tuberculosis	87
6.7	Immune reconstitution inflammatory syndrome (IRIS)	89

6.1 Testing for HIV infection in TB patients

Globally, TB is the most significant infection complicating HIV infection. In Australia, there is relatively limited overlap between those at risk of HIV infection – predominantly men who have sex with men – and those at risk of TB, who are largely people born overseas in countries of high TB incidence. TB accounts for only 2–3% of AIDS-defining illnesses in this country. However, TB is the initial AIDS-defining illness in more than 20% of locally notified AIDS patients who were born in African and Asian countries. Late diagnosis of HIV infection is common in these patients, in whom tuberculosis is often the initial manifestation of previously unsuspected HIV infection.

The natural history of TB infection is profoundly altered by HIV infection: the rate of progression from latent to active TB infection is as high as 5–10% per year in people with HIV infection, compared with 5–10% over a lifetime in latently infected individuals without HIV infection.

The overall prevalence of HIV infection in the Australian population is 0.05–0.1%. A minimum estimate of the prevalence of HIV infection among notified cases of TB is 1%. Therefore, all patients with newly diagnosed TB should be regarded in the same light as those belonging to 'traditional' HIV risk groups, and should be tested for HIV infection after appropriate discussion and provision of information about HIV testing.

6.2 Screening for latent TB infection in HIV patients

All patients with newly diagnosed HIV infection should be tested for latent TB infection. The basis for this recommendation is that the incidence of TB in people living with HIV infection in Australia (66 per 100 000) is more than ten times higher than the incidence in the general population (5.8 per 100 000).

The sensitivity of the tuberculin skin test (TST) is reduced by advancing immunosuppression, and a lower cut-off of 5 mm is used to define a positive result. Interferon gamma release assays (IGRAs) such as the QFN-GIT test are less affected by immunosuppression than the TST, and for this and other reasons are increasingly used as the preferred screening method for TB infection, as in other patient groups (see Chapter 5). The QFN-GIT test is used in the VIDS clinic, although this test also loses sensitivity and may be indeterminate (mitogen negative) with CD4 counts <100 per microlitre.

For patients with an initial negative or indeterminate test and a CD4 cell count < 200 per microlitre, US guidelines recommend repeat TB infection testing when the CD4 cell count increases above 200 per microlitre after antiretroviral (ARV) therapy is started.

All patients, especially those born overseas, should have a baseline CXR as part of the initial assessment following HIV diagnosis.

Patients with a positive IGRA or TST should undergo a thorough clinical assessment for evidence of active TB, and a chest radiograph, with further laboratory or radiological investigation if appropriate (see Chapter 5).

6.3 Treating latent TB in HIV patients

This is relatively straightforward and follows the principles outlined in Chapter 5.

- A 9-month course of INH 300 mg once daily is given. The usual alternative to INH is a 4-month course of rifampicin but significant interactions occur between rifampicin and some antiretroviral agents (see below). If rifampicin cannot be used, an alternative is rifabutin, with appropriate dose modification of rifabutin and interacting antiretroviral agents.
- INH hepatotoxicity is more common in HIV patients. LFTs should be monitored monthly, and patients counselled about symptoms of hepatitis.
- Pyridoxine 25 mg daily should be prescribed routinely
- INH should not be started at the same time as drugs with overlapping toxicity – there is less urgency about starting latent than active TB treatment, and it is best to wait at least 4 weeks after starting ARV therapy or

cotrimoxazole in order to avoid the problem of trying to establish which drug caused side-effects such as skin rash or hepatitis.

A common clinical dilemma is the HIV patient from a high TB incidence country with a CD4 cell count < 200 per microlitre, a negative QFN-GIT or TST and a normal CXR: should this patient be treated for latent TB, given the possibility that the screening test may be falsely negative?

Studies of latent TB treatment in HIV infection, including those undertaken in high TB incidence countries, indicate that the greatest treatment benefit occurs in those with a positive baseline test (usually TST in these studies). Therefore presumptive treatment of 'test negative' latent infection is not generally recommended, unless there are particularly compelling epidemiological circumstances indicating a very high risk of TB infection. As indicated above, patients with a negative or indeterminate TB infection test who start antiretroviral therapy when the CD4 cell count is < 200 per microlitre should have a repeat test when the CD4 cell count increases to > 200 per microlitre.

6.4 Diagnosing active TB in HIV patients

In contrast to most other serious HIV-related opportunistic infections, some cases of HIV-associated TB (a minority) occur at CD4 cell counts > 200 per microlitre. Most of these are cases of pulmonary TB, and usually present in the 'standard' fashion with the expected respiratory and constitutional symptoms, typical X-ray findings of an upper lobe cavitating opacity, and AFB smear-positive sputum. Diagnosis is relatively straightforward.

Unfortunately, when the CD4 cell count is < 200 per microlitre, as it usually is, diagnosis of HIV-related TB is much more challenging. Patients have atypical manifestations of pulmonary infection and more than half have extrapulmonary or disseminated disease. Notable features of pulmonary TB include:

(i) non-productive or absent cough with negative sputum smears
(ii) non-classical CXR features such as mid or lower zone opacities, pleural effusion or intrathoracic lymphadenopathy. At times, the CXR may even be normal.

Any form of extrapulmonary TB may occur so, in a patient with known HIV infection, TB must be part of the differential diagnosis of a wide range of HIV-related syndromes. These include:

- unexplained fever
- meningitis
- space-occupying intracerebral lesions

- choroidal and retinal lesions
- regional lymphadenopathy
- abdominal pain and diarrhoea
- unusual forms of extrapulmonary disease, including pancreatic and splenic abscesses, and disseminated cutaneous disease.

Diagnosis of HIV-related TB usually requires a high index of suspicion. It begins with obvious clues such as the epidemiological background of the patient, followed by careful clinical assessment and standard investigations such as CXR and sputum examination. CT scanning should be undertaken early to look for evidence of disseminated pulmonary disease (specify high-resolution CT scanning), lymphadenopathy or localised organ disease (e.g. CNS tuberculoma). Tissue biopsy (preferred to fine needle aspiration) is usually necessary to diagnose extrapulmonary TB. Mycobacterial blood cultures are often positive, even in cases of apparently localised disease such as pulmonary or lymph node TB.

Molecular testing of AFB smear-positive specimens and culture isolates, preferably using primers that will detect rifampicin resistance mutations, will enable rapid identification of *M. tuberculosis* and provide information about drug susceptibility (see Chapter 3). Pending results of molecular testing, any HIV-infected patient with AFB smear-positive sputum or bronchial washings must be assumed to have TB, because AFB smear-positive MAC and other non-tuberculous mycobacterial respiratory infections are uncommon in HIV-infected patients.

6.5 Treating active TB in HIV patients
6.5.1 The role of empiric TB therapy

Mortality is high in the 1 to 2 months following presentation among HIV-infected TB patients with a CD4 cell count < 200 per microlitre. Therefore, the threshold for giving empiric TB treatment to these patients should be much lower than it is for other TB patients. Empiric therapy is especially indicated for patients with suspected TB meningitis or miliary TB (just as it is in non-HIV patients), but should also be considered in other situations, including those entailing varying degrees of uncertainty about the diagnosis. Examples are:

(i) unexplained fever in a patient with a low CD4 cell count from a high TB incidence country with non-specific investigation findings
(ii) a patient from a similar background with a chronic respiratory illness, non-specific pulmonary opacities and AFB-negative smears (and negative studies for other pulmonary pathogens) on bronchoscopy.

6.5.2 Drug interactions between rifampicin, rifabutin and antiretroviral agents

Rifampicin lowers blood levels of ARV agents from the protease inhibitor (PI) and (to varying degrees) non-nucleoside reverse transcriptase (NNRTI) classes by inducing cytochrome P450 (CYP) enzymes, chiefly CYP 3A4 and to a lesser extent 2C19. Rifampicin also lowers blood levels of the chemokine receptor antagonist maraviroc and the integrase inhibitor raltegravir, the latter via induction of UDP glucuronyl transferase. Rifampicin blood levels are unaltered by ARV agents. Recommendations for use of these ARV agents with rifampicin are given in Table 6.5.1.

Rifabutin is a less potent CYP inducer than rifampicin. In addition, its own levels are increased by PIs and decreased by efavirenz, but minimally affected by maraviroc and raltegravir. Recommendations for co-administration of rifabutin and relevant ARV agents are also given in Table 6.5.1.

There are no significant interactions between rifampicin or rifabutin and nucleoside/nucleotide reverse transcriptase inhibitors (NRTI), nor between any ARV agents and the other first- or second-line antituberculous drugs.

6.5.3 TB treatment regimens

TB treatment regimens are essentially the same as for non-HIV infected patients but some specific issues apply.

- Patients should be managed by a physician experienced in management of both HIV and TB. If this is not possible, close liaison between the HIV and TB treatment teams is essential.
- A rifampicin-containing regimen is preferred, but in situations where rifampicin is contraindicated because of interactions with ARV agents, rifabutin can be given (with dose modification, as above) and is of similar efficacy to rifampicin. Either rifampicin or rifabutin must be used throughout the treatment course unless the patient develops side-effects or the organism is resistant.
- Most studies suggest a higher rate of side-effects from TB drugs, such as hepatitis, rash and neuropathy, although it is not always clear whether the TB drugs themselves are responsible. Patients should be counselled appropriately and monitored closely for signs of drug toxicity.
- Patients should be reviewed 2 weeks after starting treatment then at least monthly and LFTs should be done at each visit.
- Pyridoxine should be prescribed routinely.

Table 6.5.1 Interactions and dosing with rifamycins and selected ARV agents or classes

ARV agent/class	Rifampicin		Rifabutin (RBT)		
	ARV levels	ARV dose	ARV levels	ARV dose	RBT dose
Efavirenz	↓	600–800 mg daily	↑	600 mg daily	↓ 450 mg daily
Nevirapine	↓↓	Relative contraindication	↑	200 mg bd	↑ 300 mg daily
Protease inhibitors	↓↓↓	Contraindicated	↓	Standard dosing	↑↑ 150 mg daily 3 × weekly
Etravirine	↓↓↓	Contraindicated	↓	200 mg bd	↑ 300 mg daily
Raltegravir	↓	800 mg bd	↑	400 mg bd	↑ 300 mg daily
Maraviroc	↓↓	600 mg daily	↑	300 mg daily	↑ 300 mg daily

Key:
↑ blood levels increased
→ blood levels unchanged
↓ blood levels lower

- For drug-sensitive infections, standard recommendations for duration of treatment generally apply (see Chapter 1), but in situations where a 6-month course of therapy is advised there should be a low threshold for continuing treatment for 9 months (that is, HR is given in the continuation phase for 7 rather than 4 months), particularly when clinical or microbiological response is delayed. Some experts advise that a 9-month course should be given routinely.
- Directly observed therapy is advised by WHO and other authorities. In Victoria, provision of DOT should be discussed with the DoH TB Control Program, especially for patients with suspected or documented adherence problems. If DOT is used it should be given at least 3 times per week.
- Treatment of drug-resistant infections should follow the recommendations in Chapter 3, but should be extended to cover the maximum durations specified.
- Corticosteroids should be used if the HIV-positive patient meets the standard criteria for use in TB, i.e. tuberculous pericarditis and meningitis (see Chapter 2). If the risk of immune reconstitution inflammatory syndrome (discussed subsequently) is high, corticosteroids may also be considered for other indications, such as mediastinal lymph node enlargement, ureteric TB, and spinal TB with epidural involvement.
- Education and support are essential. Being diagnosed with TB and starting treatment are stressful enough at the best of times, but when this occurs in the context of HIV-related events (e.g. recent diagnosis of HIV, ongoing HIV-related complications like diarrhoea, starting ARV therapy), the patient can feel overwhelmed. Immigration status, inability to speak English, financial difficulties and employment or study concerns can add to the complexity of these cases. Special attention must be paid to adherence, which can be a particular challenge. Intensive support from clinic staff, and involvement of the social worker, public health TB program nurses and community agencies is essential, as is use of a professional interpreter if necessary.

6.6 Antiretroviral regimens in patients being treated for tuberculosis

General issues concerning selection of an ARV regimen are covered in the Australian commentary on the US Department of Health and Human Services antiretroviral treatment guidelines for adults and adolescents, available at <www.ashm.org.au/aust-guidelines>.

Most issues regarding coadministration of TB and HIV treatment relate to the potential for drug interactions between rifampicin or rifabutin and selected antiretroviral agents (see above). The initial ARV regimen should also be chosen on the basis of results of genotypic resistance testing.

No prior ARV therapy – fully drug-susceptible HIV
The preferred ARV regimen is:

tenofovir plus emtricitabine and efavirenz

(efavirenz is contraindicated in pregnancy)

UK guidelines recommend weight-based efavirenz dosing: 600 mg daily if < 60 kg (use the fixed dose combination Atripla) and 800 mg daily if > 60 kg (use Atripla plus efavirenz 200 mg). This is considered optional in US guidelines.

TB treatment:

The standard regimen containing rifampicin should be used.

No prior ARV therapy – resistance testing results not available, or documented NNRTI resistance
The preferred ARV regimen is:

tenofovir plus FTC (Truvada) and a ritonavir-boosted PI (either lopinavir or atazanavir)

TB treatment:

A regimen containing rifabutin 150 mg three times weekly instead of rifampicin should be used.

Patient currently being treated with ritonavir-boosted PI-based regimen
The preferred approach, especially if the regimen is well tolerated and the patient has a non-detectable HIV viral load, is:

continue ritonavir-boosted PI-based regimen, and use rifabutin 150 mg three times weekly instead of rifampicin.

An alternative approach in patients with no NNRTI resistance and no prior NRTI resistance is:

switch to an efavirenz-based regimen, and use rifampicin

Patients unable to take either efavirenz or PI-based regimen
Options are:

Table 6.6.2. When to start ARV therapy in previously untreated patients being treated for TB.

CD4 cell count (per microlitre)	Timing of ARV initiation after starting TB treatment
< 100	Within 2 weeks
100–500	4–8 weeks
> 500	4–8 weeks, or wait until completion of TB treatment

- Preferred (but limited experience)

tenofovir plus emtricitabine and double-dose raltegravir (800 mg twice daily), and use rifampicin

- Alternative:

tenofovir plus emtricitabine and nevirapine (providing CD4 criteria for initiating nevirapine are met – see antiretroviral guidelines), and use rifabutin

6.6.1 When to start ARV treatment

In HIV-infected patients diagnosed with TB who are not being treated with ARV therapy, treatment of TB takes precedence and must always begin before ARV therapy. Timing of introduction of ARV therapy has to balance those factors that favour delaying treatment, such as large pill burden if all drugs are introduced over a short period of time (with potential for poor tolerability and adherence), drug toxicities (with overlapping toxicities making it difficult to attribute side-effects to a particular drug), and risk of immune reconstitution inflammatory syndrome (discussed below) against the one factor favouring early treatment, that is the well-documented increased risk of death in the first one to two months following TB diagnosis in those with a CD4 cell count < 200 per microlitre. Recommendations based on the findings of recent studies are summarised in Table 6.6.2.

6.7 Immune reconstitution inflammatory syndrome (IRIS)

Two forms of TB immune reconstitution inflammatory syndrome occur in patients who start ARV therapy.

1. *Paradoxical* IRIS is the most common form. Typical patients have been recently diagnosed with TB, have a low CD4 cell count (< 100 per microlitre)

and, within 2–6 weeks of starting ARV therapy, develop inflammatory manifestations related to TB in association with reduction in HIV viral load and restoration of immune function. These reactions also occur in non-HIV infected patients.
2. *Incident* TB IRIS refers to previously undiagnosed TB that is 'unmasked' soon after starting ARV therapy.

Clinical features of TB IRIS include fever, enlarging lymph nodes (cervical, intrathoracic or intra-abdominal), worsening pulmonary infiltrates, pleural effusion, intracerebral tuberculoma, and cold abscess formation.

The overall rate of IRIS ranges from 10 to 40%. Risk factors include low CD4 cell count (especially < 50 per microlitre), extrapulmonary or disseminated TB, initiation of ARV therapy within 4–8 weeks of initiation of TB treatment, rapid decline in HIV viral load and increase in CD4 cell count.

In vitro tests have shown that these reactions occur in association with improvements in cell-mediated immunity against *M. tuberculosis* brought about by ARV therapy.

Management of these reactions comprises symptomatic treatment initially, but a short course of corticosteroids (e.g. prednisolone 40 mg initially, weaning over 4–8 weeks) is often necessary. TB treatment should be continued. Occasional deaths have been reported but most patients recover uneventfully.

References and further reading

Centers for Disease Control and Prevention. Managing drug interactions in the treatment of HIV-related tuberculosis [online]. 2007. Available at: <www.cdc.gov/tb/TB_HIV_Drugs/default.htm>

Panel on Antiretroviral Guidelines for Adult and Adolescents. Guidelines for the use of antiretroviral agents in HIV-infected adults and adolescents. Department of Health and Human Services December 2011. Guidelines and Australian commentary available at: <www.ashm.org.au/aust-guidelines>

Nahid P, Gonzalez LC, Rudoy I, et al. Treatment outcomes of patients with HIV and tuberculosis. Am J Resp Crit Care Med 2007; 175:1199–1206.

McIlleron H, Meintjes G, Burman WJ, Maartens G. Complications of antiretroviral therapy in patients with tuberculosis: drug interactions, toxicity, and immune reconstitution inflammatory syndrome. J Infect Dis 2007; 196:S63–75.

Pozniak AL, Miller RF, Lipman MCI, et al. on behalf of the BHIVA Guidelines Committee. British HIV Association guidelines for the treatment of TB/HIV coinfection 2011. HIV Med 2011; 12:517–24.

Torok ME, Farrar JJ. When to start antiretroviral therapy in HIV-associated tuberculosis. New Engl J Med 2011; 365:1538–40.

Chapter 7

Mycobacteria other than tuberculosis

7.1	Non-tuberculous mycobacteria	93
7.2	Pulmonary *Mycobacterium avium* complex infections	94
7.3	Treatment of MAC extrapulmonary disease	102
7.4	MAC hypersensitivity-like disease	102
7.5	*Mycobacterium kansasii*	102
7.6	Rapidly growing mycobacteria (RGM)	104
7.7	Pulmonary disease due to rapidly growing mycobacteria	107
7.8	*Mycobacterium marinum*	110
7.9	*Mycobacterium ulcerans*	111

7.1 Non-tuberculous mycobacteria

About 10% of mycobacterial infections seen in clinical practice are caused not by *Mycobacterium tuberculosis* but by atypical mycobacteria. *Mycobacterium avium* complex (MAC) and other nontuberculous mycobacteria (NTM) have distinctive laboratory characteristics, occur ubiquitously in the environment, are not communicable from person to person, and are strikingly resistant to antituberculous drugs.

The incidence of NTM varies in different studies between 0.7 and 1.8 per 100 000 person-years. A recent Danish population-level survey of NTM confirms that MAC is by far the most common cause of confirmed NTM pulmonary disease (> 50%) and *M. gordonnae* the most common colonising organism. *M. abscessus* was the next most common individual organism to cause pulmonary disease, although this was much less frequent (6.9%). The 5-year mortality after confirmed NTM pulmonary disease was high (40%). This distribution of disease due to NTM species is similar in a recent report from South Korea.

In immunocompetent persons the nontuberculous mycobacterial diseases are similar to tuberculosis in many ways, but they differ in several important respects:

- the diseases tend to remain localised and progress extremely slowly
- constitutional symptoms are less prominent

- isolation of a nontuberculous mycobacterium from pulmonary secretions does not always confirm a diagnosis.

7.2 Pulmonary *Mycobacterium avium* complex infections

The lungs are the most frequent site of involvement for MAC; however, all the NTM have been associated with pulmonary disease. Disease occurs more commonly in men, and there is a definite association with pre-existing lung disease, such as silicosis, bronchiectasis, or old tuberculosis.

The diagnosis of lung disease caused by MAC is based on a combination of clinical, radiographic and bacteriologic criteria, and the exclusion of other diseases that can resemble the condition. Complementary data are important for diagnosis because NTM organisms can reside in or colonise the airways without causing clinical disease, especially in patients with AIDS, and many patients have pre-existing lung disease that may make their chest radiographs abnormal.

7.2.1 Symptoms and signs

- MAC causes a chronic, slowly progressive pulmonary infection resembling tuberculosis in immunocompetent patients, many but not all of whom have underlying pulmonary disease such as COPD, bronchiectasis, previous mycobacterial disease, cystic fibrosis, and pneumoconiosis.
- Most patients with MAC infection experience a chronic cough, sputum production, and fatigue.
- Less common symptoms include malaise, dyspnoea, fever, haemoptysis, and weight loss.
- Symptoms from the underlying lung disease may confound the evaluation.
- Physical findings include fever and altered breath sounds, including rales or rhonchi, as well as signs of the underlying lung disease if present.

7.2.2 Clinical and radiological patterns of MAC lung disease

There are two typical clinical and radiological patterns of MAC lung disease in HIV-negative patients.

Fibrocavitary disease

The most common pattern of MAC lung disease is upper-lobe cavitary lung disease with clinical and radiological features that mimic TB. The cavities, however, are often thin-walled and have less surrounding parenchymal infiltrate but more prominent pleural involvement than is commonly seen with TB. These patients are predominantly older (commencing in late 40s–50s) male smokers with coexistent COPD or another underlying pulmonary disease. There may be a history of alcoholism. Patients present with cough, weight loss +/– haemoptysis. If left untreated, this form of disease is generally progressive within a relatively short time frame, 1 to 2 years, and can result in extensive cavitary lung destruction and respiratory failure. MAC disease is regarded as a superinfection.

Nodular/bronchiectatic disease
These patients are predominantly nonsmoking women over the age of 50 who do not have underlying lung disease, and have non-cavitary radiographic changes. They often have particular physical characteristics – a thin body habitus, sometimes with scoliosis, pectus excavatum, mitral valve prolapse, and joint hypermobility. High-resolution CT (HRCT) scanning reveals (i) multifocal bronchiectasis and (ii) small, predominantly peripheral, nodular densities exhibiting a 'tree-in-bud' appearance that reflects inflammatory changes, including endobronchial spread with bronchiolitis. The disease is often focused in the lingula and right middle lobe and in the lower zones. This combination of findings is highly suggestive of MAC infection. In the absence of any past history, it is unclear whether bronchiectasis actually predates and predisposes to MAC infection, although once established, MAC infection clearly contributes to progression of bronchiectasis. This form of MAC infection may progress, albeit far more slowly than the fibrocavitary form, to respiratory failure.

Shedding of MAC into respiratory secretions in these patients is less consistent than in the fibrocavitary form of the disease. Sputum may be intermittently positive and/or positive with low numbers of organisms. Bronchoscopy should be done if clinical suspicion of MAC infection is high in these patients.

7.2.3 Other forms of MAC disease
MAC may occur as a superinfection in patients with brochiectasis resulting from cystic fibrosis (CF), prior bacterial or viral infections, or previously treated TB. Patients tend to be > 50 years old (other than those with CF), of either sex, and there is no apparent relationship to smoking or smoke-related lung disease. They may have either no or minimal symptoms.

TFN-α inhibitor therapy is emerging as a risk factor for MAC infection as for TB. Patients with proven MAC pulmonary disease require therapy for this prior to commencing TNF-α inhibitor therapy.

HIV-infected patients with disseminated MAC infection uncommonly present with pulmonary involvement. When it is present it most typically presents as miliary disease.

7.2.4 Diagnostic criteria

A combination of findings is required to diagnose pulmonary MAC disease. Sputum cultures positive for atypical mycobacteria do not in themselves prove infection because NTM may exist as saprophytes in the airways or as environmental contaminants. All of the following criteria must be met before pulmonary disease can be ascribed to MAC. These criteria also apply to pulmonary disease caused by other nontuberculous mycobacteria.

Clinical criteria
- There must be a compatible clinical illness: chronic cough, sputum production, and fatigue; less commonly: malaise, dyspnoea, fever, haemoptysis, and weight loss.

Radiological criteria
- There must be a compatible pulmonary process visible radiologically: nodular or cavitary opacities on chest radiograph, or an HRCT scan that shows multifocal bronchiectasis with multiple small nodules.

Microbiologic criteria
A significant isolate of MAC must be obtained from respiratory secretions or lung tissue.

A significant isolate is defined as:

- positive culture results from at least two separate sputum samples

or

- positive culture results from at least one bronchoscopy specimen – bronchial washings or BAL

or (less commonly)

- lung biopsy with granulomatous inflammation or visible AFB and positive culture for MAC, or biopsy showing indicative histopathology and respiratory cultures positive for MAC. This is more likely to be present in HIV-infected patients investigated with bronchoscopy for a pulmonary infiltrate.

Exclusion of an alternative diagnosis
There must be reasonable exclusion of other diseases that could explain the condition, in particular tuberculosis.

Patients who are suspected of having MAC lung disease but who do not meet the diagnostic criteria should be followed until the diagnosis is firmly established or excluded.

7.2.6 Issues to consider in determining the significance of a respiratory isolate of MAC

Since MAC is present in the environment and water, a single isolation of MAC by culture from sputum may be a contaminant; the same may even apply with a single isolate from a bronchoscopy specimen, if not accompanied by typical clinical and radiological manifestations, because MAC and other NTM may contaminate bronchoscopes.

In general, multiple isolates are needed from non-sterile sites to establish disease, whereas one positive culture from a sterile site, particularly where there is supportive histopathology, is usually sufficient.

In establishing a diagnosis of MAC, the clinical presentation and any predisposing factors are also helpful. Patients with pre-existing lung disease or impaired immunity (especially cell-mediated immunity) are more prone to these infections than those without such predisposing conditions.

In patients with a more chronic presentation and a radiograph that is difficult to interpret, the diagnosis of disease, as opposed to colonisation of previously damaged lung, may be difficult to make. Indeed, sometimes disease may develop after a period of colonisation.

Making the distinction between 'colonisation' and 'disease' is not as simple as it once was. With the wider use of high resolution chest CT scanning, it is clear that many patients previously considered to be 'colonised' on the basis of a normal CXR and stable symptoms actually have slowly progressive lung disease.

7.2.7 Management

Because therapy of MAC lung disease is often poorly tolerated and is not always effective, there will be patients who have MAC isolated from a respiratory specimen but who might not benefit from medical therapy.

Patients for whom treatment may be withheld
- Patients with a normal CXR do not require therapy if they have transient or self-limited respiratory symptoms with a single smear-negative, MAC culture-positive sputum specimen, who on follow-up have multiple smear- and culture-negative specimens. They should have ongoing clini-

cal follow-up, including further CXR, to confirm the lack of significant lung disease.
- Certain patients with established MAC lung disease might not benefit from MAC medications. These include patients who have severe co-morbidities that will limit life expectancy.
- A few patients who have significant hypersensitivity responses to either macrolides or rifampicin should not receive these drugs.

Patients for whom treatment is indicated
- Patients who meet previously described diagnostic criteria for pulmonary MAC infection and who have no contraindications to anti-MAC medications.

Patients for whom treatment decision is to be deliberated.
- A risk–benefit approach needs to be considered in patients with slowly progressive disease who demonstrate intolerance of therapy or who are likely to be intolerant of therapy, as side-effects can manifestly reduce the quality of a patient's life.
- Some elderly patients with non-cavitary MAC lung disease will have slowly progressive disease that is not especially bothersome from a symptomatic standpoint and is not likely to affect life expectancy. These patients require assessment over a prolonged period in order to be sure that they have indolent disease and that they have few and relatively stable symptoms due to MAC disease. If there is acceleration of the patient's MAC disease, on either a symptomatic, microbiological or radiological basis, then the decision to withhold treatment should be revisited.

Patients for whom treatment of MAC lung disease may be modified
- Some patients, especially elderly patients, will simply decide that the adverse effects of the medication are less tolerable than the symptoms associated with MAC lung disease, regardless of the symptoms associated with MAC disease, or the rate of progression or the extent of the disease on CXR.
- If patients on a standard MAC treatment regimen (discussed in more detail below) develop gastrointestinal side-effects, it may be possible to continue therapy with a modified regimen by omitting rifampicin and using a standard or a lower dose of the macrolide agent and continuing ethambutol. Although the efficacy of this approach is unproven in MAC lung disease, clarithromycin (given at standard doses) and ethambutol is an effective regimen for disseminated MAC disease in HIV infection.

7.2.8 Treatment of MAC lung disease

VIDS generally follows the ATS recommendations on drug treatment of MAC disease. These utilise the newer macrolides as the cornerstone of therapy. The ATS guidelines recommend macrolide susceptibility testing for all new, previously untreated MAC isolates, but currently this is not standard practice in Victoria; instead such testing is reserved for those with recurrent disease.

The preferred empirical combination regimens of the ATS include a macrolide agent (clarithromycin or azithromycin) plus ethambutol plus rifampicin. With these regimens sputum conversion rates for pulmonary MAC infection in adults able to tolerate the medications are about 90%. Rifampicin is preferred over rifabutin because there is no difference in clinical response between these agents and rifabutin may cause more problematic, treatment-limiting, adverse effects (uveitis, leucopenia), especially in the elderly.

American Thoracic Society recommendation for MAC disease (Griffith et al. 2007)
For fibrocavitary disease
Clarithromycin 500 mg twice daily

or

Azithromycin 250 mg daily

plus

Ethambutol 15 mg/kg daily

plus

Rifampicin 600 mg daily

The ATS also discuss the option of adding amikacin (25 mg/kg two to three times a week) as a fourth drug for the first 8 weeks for patients who have a substantial burden of extracellular organisms, e.g. extensive or cavitary disease with strongly positive sputum smears. (Streptomycin has been used in the past but its unavailability limits current use.) We rarely follow this approach in VIDS, although it could be considered in the very occasional patient who is critically ill, or as part of a retreatment regimen.

Intermittent dosing may be used for patients with **nodular/bronchiectatic disease** consisting of:

Clarithromycin, 1000 mg three times a week

or

Azithromycin, 500 mg three times a week

plus

Ethambutol, 25 mg/kg three times a week

plus

Rifampicin, 600 mg three times a week

Lower doses of clarithromycin (500 mg/day) or azithromycin (250 mg three times a week) may be better tolerated in patients over the age of 70 years who have a low body mass, especially if the creatinine clearance is reduced. In these patients, regular doses may lead to high drug levels which may be associated with intolerable side-effects.

The duration of treatment is typically 18–24 months total with treatment continued for 12 months after monthly sputum cultures convert to negative. Most patients do not clear MAC from their sputum for between 6 and 12 months. Patients in whom sputum cultures remain positive at 9–12 months must be assessed for non-compliance.

Progression of pulmonary infiltrates during therapy or lack of radiographic improvement over time are poor prognostic signs and also raise concerns about secondary or alternative pulmonary processes.

Patients need to be reviewed monthly with repeat sputum samples done until cultures are negative. CXRs should be done at regular intervals (3-monthly). Clearing of pulmonary infiltrates due to MAC is slow.

As patients will be on long-term rifampicin and ethambutol it is important that special attention is paid to side-effects of these agents. Regular monitoring of LFTs is required for rifampicin toxicity (monthly initially then 2-monthly if well tolerated). Visual acuity and colour vision should be measured at each monthly visit, and the patient referred to the ophthalmology clinic if abnormalities are present at baseline or develop during ethambutol therapy or if the patient complains of visual symptoms.

The BTS's view appears to differ from the ATS. The 1999 BTS recommendation for HIV-negative patients with disease due to MAC is that first-line treatment should be with rifampicin and ethambutol for 24 months, plus or minus isoniazid. The British opinion is also presented here as it is useful on occasions when macrolides cannot be used. With this regimen there was 72% microbiological clearance and disease-free follow-up at 3 years.

Table 7.2.1 Adverse events associated with medications for MAC

Medication	Adverse event
Clarithromycin and azithromycin	Bitter taste, diarrhoea, anorexia, nausea, vomiting, abnormal LFTs with hepatitic picture, decreased auditory acuity, hypersensitivity reactions
Rifampicin	Nausea, vomiting, anorexia, abnormal hepatic enzymes, hypersensitivity reactions, flu-like syndrome, thrombocytopenia, renal failure
Ethambutol	Optic neuritis with loss of red-green colour discrimination and/or loss of visual acuity
Rifabutin	Nausea, vomiting, anorexia, abnormal LFTs with hepatitic picture, polyarthralgia, polymyalgia, hyperpigmentation, leucopenia, anterior uveitis, hypersensitivity reactions (thrombocytopenia, renal impairment, fever)

7.2.9 Points relevant in risk-benefit considerations for treatment of MAC lung disease

- Macrolides (clarithromycin, azithromycin) should not be used as monotherapy because of the risk of developing macrolide-resistant MAC isolates.
- The toxicity of clarithromycin and azithromycin is related to dosage and serum concentration. Doses greater than clarithromycin 1000 mg/day and azithromycin 300 mg/day are poorly tolerated.
- The regimen with the highest sputum conversion rates comprises daily clarithromycin, a rifamycin agent (VIDS follows ATS recommendations and uses rifampicin) and ethambutol.
- In multidrug regimens for MAC lung disease, dosage adjustments are frequently necessary because of adverse events.
- Close toxicity monitoring is required for all patients receiving treatment for MAC lung disease.
- In patients with localised pulmonary disease, resective lung surgery with lobectomy should be considered when there is failure of sputum conversion after 4–6 months of antimycobacterial therapy.
- Where macrolide resistance is present as in relapsed disease, the regimen most frequently used consists of rifampicin, ethambutol, moxifloxacin, and amikacin. There are limited data regarding the benefit of moxifloxacin other than animal models, but few other options are available. As

there is cross-resistance between the macrolides there is no benefit in switching between them in this situation.
- If patients are intolerant of an initial daily regimen, intermittent therapy can be tried with clarithromycin, 1000 mg; ethambutol, 25 mg/kg and rifampicin 600 mg, all given three times weekly. It may also be worthwhile changing from clarithromycin to azithromycin or vice versa in a daily regimen to rule out intolerance to a specific macrolide.

7.3 Treatment of MAC extrapulmonary disease

MAC extrapulmonary disease occurs predominantly in the cervical lymph nodes of children < 3 years of age. The treatment of choice is complete excision of the affected nodes. Antimycobacterial chemotherapy with clarithromycin, ethambutol and rifampicin for up to 2 years should be considered in those patients where disease recurs or where surgical excision is incomplete or impossible because of involvement or proximity of vital structures. The default diagnosis in a child from a TB-endemic area with granulomatous inflammation of a lymph node should be TB. In sites other than lymph nodes ATS recommends chemotherapy for 6–12 months with surgical excision for localised disease.

7.4 MAC hypersensitivity-like disease

A MAC pulmonary disease syndrome with a presentation similar to hypersensitivity lung disease has recently been recognised. This syndrome has previously been termed 'hot tub lung'. Non-tuberculous mycobacteria other than MAC also have the potential to result in this condition. MAC, like other mycobacterial organisms, has a predisposition for growth in indoor hot tubs as these bacteria are relatively resistant to disinfectants and may be able to grow at high temperatures.

Treatment requires recognition of the condition and avoidance of exposure to the hot tub. Maintenance and disinfection of the spa according to the manufacturer's instructions may be insufficient to eradicate MAC. Special disinfection measures may be successful, but most experts advise affected patients to avoid exposure to the spa completely. The role of short courses of MAC therapy and corticosteroids is not proven.

7.5 *Mycobacterium kansasii*

7.5.1 Characteristics
- Pulmonary disease is the most frequent clinical presentation of *M. kansasii* infection. Most patients are middle-aged to elderly men, over half of whom have chronic bronchitis and emphysema, old healed tuberculosis, or both.

- *M. kansasii* is found in tap water. Pulmonary disease is not transmissible from person to person.
- *M. kansasii* isolated from respiratory cultures is generally regarded as a pathogen. Isolation of *M. kansasii* from respiratory cultures is usually correlated with invasive pulmonary disease, unlike most other NTM.
- Unlike MAC, *M. kansasii* primarily causes pulmonary disease, only occasionally disseminating in both immunocompromised and immunocompetent hosts.
- The illness presents acutely or subacutely in a way that is clinically and radiologically similar to TB, although patients may be asymptomatic.
- The characteristic radiological presentation of thin-walled cavitary lesions (70–90%) is seen in both HIV and non-HIV patients.
- Less frequently, *M. kansasii* can produce non-cavitary, nodular/bronchiectatic disease similar to MAC.
- Culture techniques and drug susceptibility testing similar to that used with *M. tuberculosis* can also be performed in patients with *M. kansasii* infection.
- In untreated patients with pulmonary disease caused by *M. kansasii*, sputum positivity generally persists and the disease progresses clinically and radiologically. Clinical data suggest that patients do better with earlier therapy.
- *M. kansasii* produces proteins including CFP-10 and ESAT-6 that are detected by IGRA testing, thus a false positive test for *M. tuberculosis* infection may result.

7.5.2 Treatment

Recommended treatment of *M. kansasii* infection, with or without concurrent HIV infection (ATS recommendation 2007) is a combination of the following agents:

Isoniazid (INH) 300 mg daily

Rifampicin 600 mg daily

Ethambutol 15 mg/kg daily

- Duration of treatment is at least 12 months following negative cultures.
- This treatment regimen is the same for both extra-pulmonary and pulmonary disease.
- For patients with rifampicin-resistant *M. kansasii* disease, a three-drug regimen is recommended based on *in vitro* susceptibilities including

clarithromycin or azithromycin, moxifloxacin, ethambutol, sulfamethoxazole, or amikacin.
- Pyrazinamide is not used to treat *M. kansasii* infections because this organism is resistant to this agent.

7.6 Rapidly growing mycobacteria (RGM)

RGM are mycobacteria that produce visible non-pigmented colonies by 7 days when subcultured onto growth media. They will often have grown by 3–5 days but may take longer to grow during primary isolation.

RGM are ubiquitous in the environment, and there are approximately 50 species. The three clinically relevant species are *M. fortuitum*, *M. abscessus* (formerly *M. chelonae* subsp. *abscessus*) and *M. chelonae*. Other potentially pathogenic species include *M. smegmatis*, *M. peregrinum* and *M. chelonae*-like organisms, but these rarely cause human disease.

RGM grow well on both standard mycobacterial media and routine bacterial media (e.g. 5% sheep's blood or horse blood agar). In cutaneous disease RGM are isolated in more than 50% of cases via routine bacterial media. Nevertheless, in suspected cases, specimens should be inoculated onto mycobacterial media, and in addition cultured at both 30°C and 37°C, as some strains require lower temperatures to promote growth. Further species identification is achieved via biochemical methods and drug susceptibility patterns. *M. abscessus* and *M. chelonae* differ by only a small number of base-pairs.

RGM predominantly infest skin/soft tissue and contiguous structures, e.g. bony infection. They less commonly cause pulmonary infection.

7.6.1 *M. fortuitum*

M. fortuitum causes human infection primarily by direct inoculation, including primary skin and soft tissue infections, surgical wound infections, and catheter-related sepsis. Rarely, other infections occur such as keratitis, pulmonary disease, prosthetic valve endocarditis, and cervical lymphadenitis. *M. fortuitum* causes less than 20% of lung disease due to RGM.

7.6.2 *M. abscessus*

There is a wide clinical spectrum of disease due to *M. abscessus*. It is most frequently associated with cutaneous disease. This may be primary, such as following soil- or water-inoculation injury, or nosocomial, involving surgical wound infections. Other nosocomial infections include post-injection site abscesses, bone and joint disease, prosthetic valve endocarditis, and keratitis. It is an uncommon but

Table 7.6.1 Published susceptibilities for RGM (expressed as percentage sensitive)

Drug	M. abscessus	M. chelonae	M. fortuitum
Amikacin	90	50	100
Cefoxitin	70	0	50
Imipenem	50	60	100
Tobramycin	60	100	
Clarithromycin	100	100	80
Doxycycline	4	25	50
Ciprofloxacin	0	0	100
Sulfamethoxazole	0	3	100
Linezolid	Some activity	90	

well recognised cause of pulmonary disease. It is responsible for over 80% of RGM lung disease, but accounts for less than 5% of NTM pulmonary disease.

Disseminated infections occur, but usually in the setting of immune compromise. About 20% of cases developed in people without an identified immune defect in one review.

7.6.3 *M. chelonae*
Disseminated skin lesions involving the lower extremities (pseudo erythema nodosum) due to RGM are almost always due to *M. chelonae* and occur in patients on chronic corticosteroid therapy. In addition, surgical wound infections following augmentation mammoplasty and heart surgery have been described. *M. chelonae* is an extremely uncommon cause of RGM pulmonary disease (< 1%).

7.6.4 Antimicrobial susceptibility
RGM are uniformly resistant *in vitro* to the standard antituberculous drugs, and this is mirrored by clinical *in vivo* resistance. The exception is *M. smegmatis*, which is susceptible to ethambutol. Sensitivity studies should not be undertaken with antituberculous agents, but rather with the following antibacterial agents:

- amikacin
- cefoxitin
- imipenem
- doxycycline
- clarithromycin

- erythromycin
- ciprofloxacin/moxifloxacin
- sulfamethoxazole
- linezolid.

7.6.5 Skin and contiguous structure disease due to RGM
- Some minor infections will resolve spontaneously or after surgical debridement. However, several studies of injection site abscesses in which no therapy was given revealed disease that persisted in most patients for 8–12 months before resolving spontaneously.
- In two outbreaks of sternal wound infections caused by *M. abscessus* in the era when little was known of chemotherapy or surgery for these organisms, approximately one-third of the patients died of uncontrolled infection. Drug therapy or combined surgical and medical therapy clearly produces better results than these historical controls.

Treatment of skin and contiguous structure RGM diseases
- No controlled clinical trials of treatment for disease caused by *M. fortuitum*, *M. abscessus*, or *M. chelonae*, comparing one form of treatment with another or with no drug treatment at all, have been performed.
- Because of variable drug susceptibility among species and even within species and subgroups, susceptibility testing of all clinically significant isolates is essential for good patient management.
- Use at least two active agents initially. For *M. abscessus* and *M. chelonae*, an initial phase of combination parenteral therapy is usually given followed by oral therapy, whereas *M. fortuitum* infections, unless severe, can generally be treated with an oral regimen from the outset.

Parenteral therapy
- For *M. abscessus*, use a combination of amikacin 10–15 mg/kg IV daily and cefoxitin 200 mg/kg IV per day in divided doses; the lower amikacin dose (10 mg/kg) should be used in patients over the age of 50. Once-daily amikacin dosing is unproven clinically but appears to be adequate.
- For *M. chelonae* use tobramycin 4–6 mg/kg IV daily and imipenem 500 mg IV four times daily due to resistance to amikacin and cefoxitin.
- This combination is recommended for initial treatment (minimum 2 weeks) until clinical improvement is evident.

Oral therapy
- For *M. chelonae*, clarithromycin, linezolid (and in 20% of isolates ciprofloxacin and doxycycline) are the only oral drugs to which the organism is susceptible *in vitro*.
- For *M. abscessus*, the only oral agents available for treatment are clarithromycin (500 mg twice daily) and azithromycin (250 mg daily).
- Milder *M. fortuitum* infections can be treated with trimethoprim/sulfamethoxazole 1–2 double-strength tablets per day and doxycycline 200 mg per day if the organism is sensitive.

Duration of treatment
It is not certain how long chemotherapy should be continued for these infections as there is no evidence from controlled clinical trials. For serious disease, a minimum of 4 months of therapy is necessary to provide a high likelihood of cure. For bone infections, 6 months of therapy is recommended.

Surgery
Surgery is generally indicated for both extensive disease and abscess formation where drug therapy is difficult. Removal of foreign bodies such as breast implants, percutaneous catheters etc. is essential for recovery.

7.7 Pulmonary disease due to rapidly growing mycobacteria

- Pulmonary disease due to RGM is well described and most commonly occurs in the elderly or those with an underlying predisposing condition.
- The largest group of patients with this lung disease are elderly (> 60), white, female nonsmokers with no predisposing conditions or known lung disease.
- Underlying disorders that are associated with the disease include lung damage produced by prior mycobacterial infection (usually tuberculosis or MAC), gastro-oesophageal disorders with chronic vomiting, lipoid pneumonia, cystic fibrosis, and bronchiectasis due to prior respiratory infection.
- The distinguishing feature of patients with underlying disease is that their rapidly growing mycobacteria lung disease occurs at a younger age, usually < 50. Almost all patients < 40 have one of the above disorders.
- Overall, *M. abscessus* appears to be a more virulent respiratory pathogen than *M. fortuitum*.

- Careful clinical evaluation and follow-up, as for pulmonary MAC infection, is always necessary to determine the significance of an RGM respiratory isolate.
- Interestingly, approximately 15% of patients with *M. abscessus* will also have *M. avium* complex, suggesting the close relationship of the disorders. Some patients have positive sputum cultures for *Pseudomonas aeruginosa*, further evidence of bronchiectasis.

7.7.1 Clinical features
- The usual presenting symptoms are cough and easy fatiguability, often attributed for months or years to bronchitis or bronchiectasis.
- Fever, night sweats, and weight loss occur, but they are much less common and less severe than with *M. tuberculosis*.
- Haemoptysis and dyspnoea are also common.
- The constellation of typical presenting symptoms in an elderly nonsmoking patient with no underlying lung disease, a compatible chest radiograph, and multiple culture positive sputum specimens is sufficient to make a diagnosis.
- The presence of other diseases or unusual features may necessitate obtaining a lung biopsy to be certain of the diagnosis.
- The diagnosis of RGM pneumonia tends to be delayed. (It was usually not established until > 2 years after the onset of symptoms in one series.)

7.7.2 Natural history
- The natural history of this disease depends primarily on the presence or absence of underlying disorders.
- For most patients with *M. abscessus* and no underlying disorder, the disease is indolent and slowly progressive. Some patients show little radiographic change over years.
- More fulminant, rapidly progressive disease can occur, particularly in association with gastro-oesophageal disorders. Death occurs as a consequence of *M. abscessus* in 20% of cases.

7.7.3 Imaging
- In patients with no apparent risk factors, the chest radiograph usually shows multilobar, patchy, reticulonodular or mixed interstitial–alveolar infiltrates with upper lobe predominance.

- Cavitation occurs in only approximately 15% of cases.
- The CXR is usually not typical of or suggestive of reactivation pulmonary tuberculosis, which likely accounts for a delay in ordering sputum for AFB analysis and therefore a delay in diagnosis.
- HRCT of the lung frequently shows associated cylindrical bronchiectasis and multiple small (< 5 mm) nodules, a pattern also seen in nonsmokers with MAC lung disease.

7.7.4 Treatment of RGM pulmonary disease
- RGM pulmonary disease must be treated because the natural history usually involves slow progression, and, without treatment, death. Spontaneous resolutions may occur, but rarely.
- Appropriate therapy relies on accurate species identification and the results of susceptibility testing to the drugs listed above. Susceptibility testing should be performed on each clinically significant isolate as there may be interspecies and intraspecies variability.
- If surgery is possible it should be employed as surgical resection of localised disease is the only possible means of cure.

M. abscessus
- *M. abscessus* isolates are usually susceptible *in vitro* only to the parenteral agents amikacin, cefoxitin, and imipenem, and to the newer oral macrolides.
- Microbiological and clinical cure may not be possible and therapy may need to aim at symptomatic and radiological improvement.
- Prior to the availability of clarithromycin, medical therapy alone was largely unsuccessful for disease due to *M. abscessus*. Cure was only achieved in those where antibiotics were combined with localised surgical resection of the involved lung.
- Combination therapy of low-dose amikacin plus high-dose cefoxitin for 2–4 months along with clarithromycin should be used. Successful therapy consists of a 12-month period of negative sputum cultures.
- Quinolones, sulfonamides, amikacin, cefoxitin, imipenem, linezolid and newer agents such as tigecycline and telithromycin (not available in Australia) may have a place in treatment.
- Surgical resection for limited disease related to prior localised lung disease can also be curative.

M. fortuitum
- *M. fortuitum* isolates are usually susceptible to multiple oral antimicrobial agents including quinolones, doxycycline and minocycline, and sulfonamides.
- Inducible resistance to clarithromycin and azithromycin may be present so these agents should be used cautiously.
- Drug susceptibilities for this species are essential for effective therapy.
- Six to twelve months of therapy with two oral agents to which the *M. fortuitum* isolate is susceptible *in vitro* usually results in clinical cure.
- For serious *M. fortuitum* infections, a combination of intravenous amikacin 10–15 mg/kg IV per day, plus cefoxitin 200 mg/kg IV per day in divided doses is recommended in the initial phase of treatment.

7.8 *Mycobacterium marinum*
7.8.1 Disease characteristics
- *M. marinum* predominantly affects the skin and subcutaneous tissue.
- Infection is acquired after exposure to water, commonly from relatively stagnant water, such as in fish tanks. The disease present as slowly growing nodular–ulcerating lesions of the extremities. The lesions may be solitary but may ascend along the limbs, resembling sporotrichosis. The lesions may be extremely long-lived, up to 50 years.
- Diagnosis is made by biopsy of affected tissue for histopathological examination and culture of the organism.

By susceptibility testing, these isolates are:
- susceptible to rifampicin, ethambutol, clarithromycin, sulfonamides or trimethoprim–sulfamethoxazole
- susceptible or intermediately susceptible to doxycycline and minocycline
- intermediately susceptible to streptomycin
- resistant to isoniazid and pyrazinamide.

M. marinum produces proteins including CFP-10 and ESAT-6 that are detected by IGRA testing, thus a false positive test for *M. tuberculosis* infection may result. As the skin lesions of *M. marinum* are not typical of cutaneous TB there should be no reason for ordering QFN-GIT in these patients.

7.8.2 Treatment
- Superficial skin infections due to *M. marinum* may heal spontaneously, but deeper infections and bone involvement require definitive treatment.

- Surgical debridement, cryosurgery, or electro-dessication may be all that is required for small, superficial lesions.
- If the lesions are more extensive, then one of the following treatment regimens can be used:

Clarithromycin 500 mg twice daily

or

Minocycline or doxycycline 100 mg twice daily

or

Trimethoprim–sulfamethoxazole (cotrimoxazole) 160/800 mg twice daily

- For extensive disease, use a combination of:

Clarithromycin 500 mg daily

and

Ethambutol 15 mg/kg daily

- Rifampicin may also be added for osteomyelitis.
- The rate of response to antimycobacterial treatment is variable and response may not occur for up to 3 weeks. A minimum of 4 to 6 weeks of therapy should be given before considering that the patient may not be responding. Therapy is typically continued for a minimum of at least 3 months. Very longstanding lesions require a longer course of therapy.
- If a lesion is excised surgically, many clinicians provide drug coverage during the perioperative period. It is not clear if longer duration of drug treatment after surgery offers any additional advantage.

7.9 *Mycobacterium ulcerans*
7.9.1 Disease characteristics
Mycobacterium ulcerans causes skin and soft tissue infections, referred to as Bairnsdale or Daintree ulcers in Australia or Buruli ulcers (BU) internationally.

Its major virulence factor is a lipid toxin, capable of causing necrosis of fat and subcutaneous tissue. Significant morbidity is caused by chronic ulceration and potential disfigurement if left untreated.

Patients typically present with a slowly progressive skin papule that evolves to become an undermined ulcer over months. There is usually minimal pain, and an absence of systemic symptoms. A single arm or leg lesion is most common, but buttock, abdominal wall, and face or head lesions have all been reported in Australia. BU should be considered in cases of unresolving cellulitis or necrotising skin lesions. Rarely, the disease can present as acute limb swelling and oedema with or without the classic ulceration or skin lesion.

7.9.2 Distribution
In Australia, *M. ulcerans* is distributed in the environment in coastal Victoria in the Bellarine and Mornington peninsulas, and the Western Port and Gippsland Lakes areas. In northern Australia, active foci are found near Rockhampton and in Far North Queensland.

Overseas, *M. ulcerans* is endemic in sub-Saharan Africa, particularly in West Africa where point prevalence studies estimate BU rates in Ghana and Benin are similar to local rates of tuberculosis or leprosy. While more than 30 countries have reported cases of BU, in the Asia–Pacific region *M. ulcerans* is endemic in parts of Papua New Guinea and Malaysia. Molecular typing suggests that strains differ across geographic regions.

Residents of endemic areas are at greatest risk of infection; however, it is thought that visitors with only brief environmental contact can become infected.

7.9.3 Diagnosis
Microbiological diagnosis from wound swabs or tissue samples can be made if acid-fast bacilli are seen, particularly in a patient who has not travelled abroad. Polymerase chain reaction (directed at IS 2404) is highly sensitive and specific if sufficient material is obtained for analysis.

Samples should be collected deep to the undermined ulcer edge using standard cotton-tipped swabs in either dry or standard non-charcoal transport medium. Requests should be made for AFB microscopy, culture and PCR.

7.9.4 Treatment
A combination of surgical and medical therapy is usually required.

- Surgery is generally recommended to debride necrotic tissue. Small lesions may be completely cured by surgery alone. Case series evidence suggests that AFBs or granulomas present in resection margins predict treatment failure and the need for adjuvant medical therapy.
- Large lesions may require additional preoperative antibiotics for 2–4 weeks.

- Indications for combination antibiotics include:
  - resection margins with AFBs or granulomas present
  - large lesions requiring grafting
  - complex, recurrent disease
  - disease which cannot be complete debrided.

Oral-only regimes include:

Rifampicin 10 mg/kg per day up to 600 mg daily for 3 months

plus either

Clarithromycin 500 mg twice daily for 3 months

or

Ciprofloxacin 500–750 mg twice daily for 3 months

or

Moxifloxacin 400 mg daily for 3 months

7.9.5 Indications for intravenous therapy

WHO recommend oral rifampicin in conjunction with injectable streptomycin (interchangeable with amikacin in Australia). Oral therapy alone using two agents has been shown to have some success in Australian case series, but in animal models, oral-only therapy is less effective at killing *M. ulcerans*.

Intravenous amikacin (for 4–8 weeks) with oral rifampicin (for 3 months) should be considered for:

- severe or extensive disease
- when deep structures (bone, tendon, nerves, vessels) are involved
- large lesions that could not be fully resected
- major relapses
- osteomyelitis
- where trying to minimise surgery (e.g. eye or face lesions)
- initial therapy of acute oedematous disease.

Amikacin regimen
Amikacin 15 mg/kg (ideal body weight, maximum 1000 mg) intravenously daily on 5–7 days each week for 4–8 weeks

- Trough drug levels and renal, auditory and vestibular function need close monitoring in accordance with treatment guidelines described earlier for resistant *M. tuberculosis* (see Chapter 3).

Prolonged follow-up for recurrence after surgery and/or medical therapy is advised as late relapses can occur.

7.9.6 Prevention
- There is no immunity developed from previous BU infection, so re-infection is possible.
- There are no public health interventions known to date to remove *M. ulcerans* from the environment.
- Personal protection including arm and leg covering and insect repellent to avoid insect bites is recommended when outdoors in endemic areas.

References and further reading

Antibiotic Expert Group. Mycobacterial infections. In Therapeutic Guidelines: antibiotic. Version 14. Therapeutic Guidelines Limited, Melbourne, 2010, pp. 171–88.

De Groote MA, Huitt G. Infections due to rapidly growing mycobacteria. Clin Infect Dis 2006; 42(12):1756–63.

Gordon CL, Buntine JA, Hayman JA, et al. All-oral antibiotic treatment for Buruli ulcer: a report of four patients. PLoS Negl Trop Dis 2010; 4(11):e770.

Griffith DE, Aksamit T, Brown-Elliott BA, et al. An official ATS/IDSA statement: diagnosis, treatment, and prevention of nontuberculous mycobacterial diseases. Am J Respir Crit Care Med 2007; 175(4):367–416.

Johnson PD, Hayman JA, Quek TY, et al. *Mycobacterium ulcerans* Study Team. Consensus recommendations for the diagnosis, treatment and control of *Mycobacterium ulcerans* infection (Bairnsdale or Buruli ulcer) in Victoria, Australia. Med J Aust. 2007; 186(2):64–8.

Johnson PD, Stinear T, Small PL, et al. Buruli ulcer (*M. ulcerans* infection): new insights, new hope for disease control. PLoS Med 2005; 2(4):e108. Epub 2005 Apr 26. Erratum in: PLoS Med. 2005; 2(5):e173.

O'Brien DP, Hughes AJ, Cheng AC, et al. Outcomes for *Mycobacterium ulcerans* infection with combined surgery and antibiotic therapy: findings from a south-eastern Australian case series. Med J Aust. 2007; 186(2):58–61.

Subcommittee of the Joint Tuberculosis Committee of the British Thoracic Society. Management of opportunist mycobacterial infections. Joint Tuberculosis Committee guidelines 1999. Thorax 2000; 55:210–18.

Chapter 8

BCG vaccination

8.1	BCG vaccine and its efficacy	115
8.2	Recommendations for BCG vaccination	116
8.3	Contraindications to BCG	118
8.4	Practical issues in BCG administration	118
8.5	Expected response following BCG vaccination	119
8.6	'Accelerated response' and its significance	120
8.7	Management of local reactions	120
8.8	Other complications of BCG vaccination	121
8.9	BCG and bladder cancer	122
8.10	Antimycobacterial drugs and BCG	127

8.1 BCG vaccine and its efficacy

BCG (Bacille Calmette-Guérin) vaccine is a suspension of live attenuated *M. bovis*, derived from the strain developed by the Institut Pasteur and first tested in humans in 1921. BCG strains differ genetically from the original strain of *M. bovis* and from each other. Compared to *M. bovis*, all BCG strains lack the RD1 region, which contains the antigen coding genes ESAT6 and CFP10. Another deletion, RD2, occurred at the Institut Pasteur between 1927 and 1931. A third deletion, RD14, occurred between 1938 and 1961. Other deletions, duplications and polymorphisms vary between strains. These genetic differences enable the distinction of infections due to *M. bovis* from those due to BCG vaccine strains, using either culture or PCR. The genetic deletions are also associated with variations in susceptibility of BCG strains to antimycobacterial drugs and antibiotics.

The vaccine was initially administered orally, but intradermal vaccination was introduced in 1927, and percutaneous administration in 1939. Oral administration ceased in 1979. The original vaccine was distributed to a number of laboratories worldwide, with subsequent propagation of a number of strains, described by the location of their production. The BCG vaccine currently in use in Australia is the Connaught strain manufactured by Sanofi Pasteur Ltd.

BCG is protective against TB and leprosy. There has been much debate on the efficacy of BCG vaccine against TB. Since 1975, case-control studies and prospective clinical trials using different BCG strains indicated that vaccine efficacies ranged from 0 to 80%. Immunity induced by the same vaccine may vary appreciably between populations.

Vaccine-induced immunity differs in degree against different forms of tuberculosis, and in particular it may be more effective against meningitic and miliary disease than against pulmonary disease. A meta-analysis found that the average protective effect of BCG was 50% against TB infection, 78% against pulmonary and disseminated TB, 64% against TB meningitis, and 71% against death from TB. That analysis also determined that higher BCG vaccine efficacy rates were not associated with the use of particular vaccine strains. Most of the variability in results arose from geographic latitude of the study site, with greater efficacy at higher latitudes.

The important point is that BCG vaccination does not consistently prevent tuberculous infection. It primarily reduces the risk of death from tuberculous meningitis and disseminated disease in young children. The duration of efficacy wanes over time. Several studies have shown the protective effect to last up to 15 years, although it may persist beyond 50 years. Subsequent revaccination with BCG has not been shown to confer any additional protection against TB.

BCG is the most widely used vaccine in the world. While it is given to most children in the developing world, its use in the developed countries is limited to selected populations. The major objection to its use is that it makes interpretation of TST difficult. There is no correlation between the presence or strength of reactivity to tuberculin and protective effect of BCG.

8.2 Recommendations for BCG vaccination
8.2.1 Australia
In Australia BCG is recommended for:

- Aboriginal and Torres Strait Islander neonates living in regions of high incidence
- neonates born to parents with leprosy or with a family history of leprosy
- children under the age of 5 years who will be travelling to live in countries of high tuberculosis prevalence for longer than 3 months (countries with an annual incidence over 100 per 100 000 population: see <http://who.int/tb/en>)

- embalmers
- healthcare workers conducting autopsies
- healthcare workers who may be at high risk of exposure to drug-resistant cases of TB.

Victoria
Victorian Department of Health Guidelines also recommend BCG for:
- children under the age of 5 years who will be travelling to live for more than 4–6 weeks in countries of high TB prevalence
- infants and children under the age of 5 years who live in a household that includes immigrants or unscreened visitors who recently arrived from countries of high TB prevalence. This group includes infants and young children in families that travel frequently to visit or stay in the homes of relatives in countries of high TB prevalence.

Additional recommendations
BCG should also be considered for:
- children and adolescents aged less than 15 years who continue to be exposed to an index case with active smear- and/or culture-positive pulmonary TB, and who cannot be placed on isoniazid therapy
- children, adolescents and young adults to the age of 25 to 30 years who have been exposed to an index case with active MDR-pulmonary TB (where the organisms are resistant at least both rifampicin and isoniazid), after exclusion of TB infection with TST or IGRA
- persons aged over 5 years through to young adulthood who are living or travelling for extended periods (more than 2–3 months) in countries of high TB prevalence.

8.2.2 International recommendations
- US guidelines recommend BCG for a narrower range of groups than those of Australia (children with continued exposure to pulmonary TB, who cannot be removed from the infectious contact *and* cannot be given primary therapy for latent TB, and healthcare workers working in areas with a high prevalence of drug-resistant TB, without effective TB infection control procedures).
- The UK guidelines recommend BCG for a wider range of groups than do those of Australia (a wide range of children and adults with risk of social or occupational exposure to TB).

8.3 Contraindications to BCG

8.3.1 Contraindications
BCG is contraindicated for:

- individuals with TST of greater than 5 mm or positive IGRA at any time in the past
- patients who are immunocompromised by HIV infection, corticosteroids, other immunosuppressive drugs, irradiation, or malignancies involving bone marrow or lymphoid systems (because of the high risk of disseminated BCG infection in these individuals)
- individuals with a high risk of HIV infection where HIV antibody status is unknown
- individuals with any serious illness, including the malnourished
- individuals with significant fever
- individuals with generalised skin diseases (e.g. eczema, dermatitis, psoriasis)
- pregnant women and those who may become pregnant soon
- individuals who have previously had tuberculosis.

8.3.2 Defer vaccination
BCG should be deferred for the following:

- neonates with birth weight < 2.5 kg or who may be relatively malnourished
- neonates of HIV-infected mothers (until HIV infection has been excluded in the infant)
- children currently on isoniazid therapy for latent TB infection (since isoniazid may inactivate the vaccine)
- recent vaccination (within past 4 weeks) of another live vaccine such as measles-mumps-rubella, varicella, or yellow fever (but **not** oral polio vaccine). However, BCG can be given **concurrently** with another live vaccine.

8.4 Practical issues in BCG administration
A TST must be carried out before BCG immunisation except in infants < 6 months old who may be immunised without a prior test unless exposure to infectious TB has occurred. If exposure to infectious TB has taken place, such infants should be tested and considered for preventive therapy rather than BCG.

BCG should not be given to individuals with a TST reaction ≥ 5 mm.

BCG can be given at the same time as live viral vaccines, measles-mumps-rubella or yellow fever. If it is not done at that time, it is necessary to wait 4–6 weeks after these live vaccines before administering BCG, as the immunogenicity of BCG may be impaired if it is given earlier. Oral polio vaccine does not affect the immunogenicity of BCG because the oral polio vaccine virus replicates in the intestine to induce local immunity and serum antibodies, and three doses are given.

No further immunisation should be given for at least 3 months in the arm used for BCG vaccination because of the risk of regional lymphadenitis.

BCG vaccine can be given only by an accredited vaccinator. The names and addresses of accredited BCG vaccinators in Victoria can be obtained from the Tuberculosis Control Section, Department of Health (telephone 03 9096 5114).

BCG is administered intradermally into the upper arm in the region of the insertion of the deltoid into the humerus. By convention, the left arm is used in order to enable subsequent identification of individuals who have received BCG vaccination. **The BCG vaccine must not be given subcutaneously.**

Protective eye wear should be worn by the person administering the vaccine, the patient, and the parent holding the patient (if the patient is a small child requiring restraint). Eye splashes may ulcerate. If an eye splash occurs, wash the eye immediately with saline or water and seek expert advice.

Dose: 0.1 mL for children and adults over 12 months old; 0.05 mL for infants under 12 months of age.

8.5 Expected response following BCG vaccination

The normal local response to BCG vaccination occurs within 2–3 weeks. Induration with erythema appears, and a small papule (5–6 mm diameter) is usually present after 2 weeks. This is followed by a pustule or an ulcer with scab. Local reaction is maximum at 4–6 weeks. Tuberculin reactivity appears at 6–10 weeks in 95% of BCG vaccine recipients.

In the great majority resolution takes place after 3–4 months leaving a scar of 4–8 mm, although up to 25% of children may not have a typical scar after vaccination in infancy. In 50,000 individuals vaccinated with BCG, less than 2% had an ulcer of 10 mm or more when examined 6–14 weeks after vaccination. In 15,000 individuals examined up to 6 months later, only 0–0.56% had ulcers larger than 10 mm.

Survival of living BCG in tissues has been demonstrated as long as 517 days after inoculation.

8.6 'Accelerated response' and its significance

When tuberculin reactors or those in the pre-allergic phase of TB infection are vaccinated a pronounced reaction frequently occurs at the site of vaccination. A nodule with induration is formed on the 1st or 2nd days (within 72 hours). Scab formation and healing may be completed by 10–15 days. The intensity of the reaction, which represents Koch's phenomenon, varies between individuals and vaccine strains.

8.7 Management of local reactions

Local reactions occur in approximately 5% of vaccinated individuals. Abscesses at the primary inoculation site occur in about 2.5%, lymphadenitis in about 1%, and severe local reactions in about 1.5% of vaccinated individuals. Erythema nodosum occurs occasionally. About 1% of individuals require medical attention (including surgery) following vaccination. Reactions are more severe when the vaccine is injected subcutaneously.

Lymphadenitis may appear early (within 2 months of vaccination) and late (between 2 and 8 months following vaccination). Lymphadenitis with 'softening' (fluctuance with abscess formation) does not usually appear before the 3rd month, sometimes as late as 6th or 7th month and exceptionally after 2–3 years. Small children are more liable to lymphadenitis with softening than older children and adults. If the enlargement is rapid (within 2 months) spontaneous drainage is more likely to occur compared to the slowly progressive form. Abscesses in lymph nodes do not always coincide with large reactions at the inoculation site.

The management of local ulceration and abscess formation at the inoculation site is controversial. In the vast majority, follow-up without specific treatment is all that is required, and this should be the mainstay of the management in BCG-related local reactions. Non-suppurative lymph nodes usually improve spontaneously, although resolution may take several months.

8.7.1 Treatment modalities

If specific treatment is required, options for medical treatment are:

- oral erythromycin 500 mg 4 times daily for 3–4 weeks
- isoniazid
- isoniazid plus rifampicin.

Medical treatments may speed the resolution of inoculation site reactions and lymphadenitis, but none of these regimens has clearly demonstrable efficacy over another, or in preventing progression to local abscess formation.

Options for surgical treatment (which may be combined with medical therapy) are:

- surgical incision and drainage – with or without installation of isoniazid into the abscess cavity (50 mg single dose in 0.5 mL solution)
- surgical drainage and excision of abscess cavity.

8.7.2 General approach
- Recognise that ulcers may take 2–3 months to heal. Reassure the patient. Take culture for secondary invaders (rare), and for BCG if the ulcer is long-lasting.
- Leave the ulcer uncovered and exposed to air. If it is moist, dab the ulcer gently with cotton wool soaked in methylated spirit, as often as required, to dry the lesion.
- Use dry dressing over the ulcer if there is a discharge.
- Patients with large discharging ulcers beyond 2–3 months, especially if associated with lymphadenitis, are likely to demand that something be done. Oral erythromycin for up to 4 weeks is a reasonable first choice. (No work has been done with the newer macrolides, but they can be used.)
- Non-adherent lymphadenitis will heal spontaneously without treatment. For adherent or fistulated lymph nodes the WHO suggests drainage and direct instillation of antituberculous drug into the lesion.
- If the node is very large (greater than 30 mm in diameter or enlarging rapidly), total excision is the treatment of choice because the recurrence rate after incision and drainage is high, and scarring is often prominent.
- If the response is an accelerated one (see section 8.6), first exclude presence of active disease. Then consider the use of isoniazid preventive therapy for *M. tuberculosis* infection. The exaggerated response itself is not expected to respond to isoniazid.

8.8 Other complications of BCG vaccination

Disseminated BCG infection is very rare, but has a case fatality rate of up to 70% despite antituberculous therapy. The incidence of fatal disseminated disease is estimated at 0.19 to 1.56 per million vaccines, and has occurred almost exclusively in those with severely compromised cellular immunity. Disseminated infection may occur decades after the vaccination. One 31-year-old HIV-infected person developed disseminated BCG infection 30 years after BCG vaccination. Treatment as for disseminated TB may be successful in some cases, noting that BCG (like other strains of *M. bovis*) is not susceptible to pyrazinamide.

BCG in patients on corticosteroid therapy may result in an excessively large BCG lesion with regional adenopathy and failure to develop tuberculin sensitivity. Hypertrophic scars occur in an estimated 28–33% of vaccinated persons taking corticosteroids, and keloid scars occur in approximately 2–4%.

The incidence of osteitis varies from 0.01 per million in Japan to 300 per million in Finland.

8.9 BCG and bladder cancer

Intravesical installation of BCG is used in the adjuvant treatment of intermediate and high-risk non-muscle-invasive bladder cancer, to reduce the risk of local recurrence and disease progression.

Intravesical installation of BCG suspension leads to a delayed-type hypersensitivity reaction in the bladder wall, but the mechanism by which BCG therapy acts against tumour recurrence and progression is not well understood.

Lower urinary tract symptoms such as dysuria, frequency, urgency, haematuria and suprapubic pain are common with intravesical BCG therapy but the severity of these symptoms does not correlate with efficacy.

Acid-fast bacilli may be visible or detectable by PCR in the bladder wall several years after installation of BCG, but viable organisms cannot usually be cultured from urine after a week or two. Mycobacteraemia (detected by PCR) following intravesical BCG is uncommon in patients without clinical side-effects.

8.9.1 Dosage and administration

Intravesical BCG is given according to a standard protocol. The original protocol used by Morales et al. consisted of 120 mg of BCG-Pasteur suspended in 50 mL of normal saline. The dose was repeated weekly for 6 weeks. The need for a maintenance phase of therapy was subsequently demonstrated by Lamm et al. and the Southwest Oncology Group. The optimum dose and frequency of maintenance treatment have not yet been established, but a minimum of three cycles (each cycle comprising weekly installations for 3 weeks) at 3, 6 and 12 months is currently thought necessary.

Induction therapy with intravesical BCG should commence a minimum of 2 weeks following transurethral resection of bladder tumour, to allow re-epithelialisation of the bladder and reduce the risk of systemic side-effects.

Indications to defer treatment
Intravesical BCG should be deferred in the presence of:

- traumatic catheterisation (defer 1 week)
- gross haematuria (defer until urine clears)
- bacterial urinary tract infection (defer 1 week, until treated)
- symptomatic cystitis (depending on severity)
- contracted bladder (defer until resolution of symptoms)
- local side-effects, i.e. symptomatic granulomatous prostatitis, epididymo-orchitis (suspend installations until treated)
- mild or transient fever and malaise (≤ 48 hours)
- mild allergic reactions (defer until resolution of symptoms).

Indications to discontinue treatment
Intravesical BCG should be permanently discontinued in the presence of:

- persistent high-grade fever (> 38.5°C for > 48 hours)
- systemic BCG reaction
- severe allergic reaction.

8.9.2 Side-effects of intravesical BCG therapy of bladder cancer
Side-effects of intravesical BCG are usually seen during induction therapy and in the first 6 months of maintenance therapy.

Cystitis
Cystitis is the most common side-effect, occurring in 80% of patients. It is frequently accompanied by haematuria and is the most common reason for delayed installation of therapy. The severity of lower urinary tract symptoms does not correlate with the efficacy of BCG therapy against tumour recurrence.

Initial management should include urine culture to exclude bacterial cystitis, and symptomatic treatment. If symptoms persist beyond 48 hours, the next installation should be postponed. If a bacterial infection is identified, treat according to culture and sensitivity testing results. Note that chemical cystitis due to intravesical chemotherapy is also common, in which case antimicrobial therapy will be of no benefit.

For subsequent installations, doses of BCG may be reduced and use of peri-installation antibacterial prophylaxis directed against coliform organisms can be considered.

Haematuria
Gross haematuria occurs in up to 90% of patients and frequently occurs with cystitis. Management includes culture of urine to exclude bacterial cystitis and postponement of intravesical therapy until urine clears. Persistent gross haematuria (> 48 h) may warrant further investigation, e.g. cystoscopy to exclude tumour recurrence, and treatment, e.g. bladder irrigation. Haematuria is also commonly due to intravesical chemotherapy.

Granulomatous prostatitis
Granulomatous prostatitis is common, but only 1–3% of patients have local or systemic symptoms. Up to 5% of cases will need treatment. Digital rectal examination may reveal an indurated prostate. Prostate specific antigen may be elevated. Ultrasound may show hypoechoic areas in the transition zone of prostate.

Initial management includes culture of urine to check for bacterial cystitis and/or prostatitis.

If symptoms are severe or persist beyond 48 hours, specific therapy is indicated, using antimycobacterial drugs for 3 months. The suggested initial regimen is:

Rifampicin 10 mg/kg (up to 600 mg) orally, daily

plus

Isoniazid 10 mg/kg (up to 300 mg) orally, daily

Corticosteroids are often used initially, and their dose is tapered once symptoms improve. An antibiotic active against coliforms may be added if bacterial infection cannot be excluded.

Response to treatment may be monitored clinically, with PSA and ultrasound. Failure of biochemical and radiological abnormalities to resolve suggests a need to exclude prostatic malignancy. Subsequent installations should be postponed until treatment is completed and symptoms resolved.

Epididymo-orchitis

Early epididymo-orchitis is usually bacterial, while later onset suggests mycobacterial infection by BCG. Estimates of the frequency of its occurrence range from 0.2–10% of patients treated with intravesical BCG.

Early management includes urine culture for bacterial pathogens and antibiotic therapy directed against coliforms. Lack of response to antibacterial therapy suggests a need for treatment directed against BCG. Antimycobacterial therapy as for granulomatous prostatitis (in combination with an antibiotic active against coliforms if bacterial infection cannot be excluded) is recommended for a period of 3 months. Corticosteroids may be used if symptoms are severe or persist.

Severe, persistent symptoms or development of an abscess may require orchidectomy. Further intravesical therapy should be postponed until treatment is completed and symptoms resolved.

Malaise and fever

If symptoms are mild and short-lived, only symptomatic treatment is needed. If high-grade fever (> 38.5°C) and malaise are persistent > 48 h, evaluation for systemic BCG infection should be undertaken, and intravesical BCG therapy should be permanently discontinued (unless BCG has been excluded as a cause).

Disseminated BCG infection

Disseminated BCG infection is rare but life-threatening granulomatous inflammatory process. Diagnosis is often difficult. Onset may be rapid, following installation of BCG, or may occur many years after completion of therapy. The course may be indolent and subacute, or fulminant with septic shock and multi-organ failure. Although there are theoretical concerns about the risk of systemic BCG infection in immunocompromised patients receiving intravesical BCG therapy, the risk of systemic BCG infection has not been shown to be higher for individuals with haematological malignancies or those on treatment with low-dose corticosteroids. Published experience with other immunocompromised patients is limited. Symptoms may be generalised and non-specific, or related to particular organs, as are those of other disseminated mycobacterial infections. Infections of prosthetic joints have been reported, as have mycotic aneurysms, endophthalmitis, osteomyelitis and lymphadenitis. Diagnosis may be confirmed by identification of *M. bovis*-BCG in blood, urine, bone marrow or other tissues, through culture or specific PCR. Often, granulomata are seen on histology of biopsy specimens but acid-fast bacilli cannot be identified. Diagnosis in these cases is presumptive.

Management requires treatment with antimycobacterial drugs for 6 months. Bone and joint infections may need a longer course of therapy (12 months). The suggested regimen is:

Rifampicin 10 mg/kg (up to 600 mg) orally, daily

plus

Isoniazid 10 mg/kg (up to 300 mg) orally, daily

plus

Ethambutol 15 mg/kg (up to 1200 mg) orally, daily

The rationale for addition of ethambutol to HR is discussed in section 8.10. Moxifloxacin 400 mg daily should be used if ethambutol is contraindicated. High-dose corticosteroids should also be used until symptoms have resolved. Intravesical BCG therapy should be discontinued permanently.

Systemic granulomatous disease
Without isolation of *M bovis* it is difficult to distinguish clinically between a systemic BCG infection and a noninfectious granulomatous BCG reaction, which also occurs. The distinction may be somewhat artificial. Treatment requires both antimycobacterial drugs and systemic corticosteroids, with subsequent permanent discontinuation of intravesical BCG therapy.

Noninfectious systemic reactions
A seronegative and culture-negative reactive polyarthritis may occur following BCG installation. Joints most commonly affected include knees, ankles and wrists, but other joints may be involved. Conjunctivitis or uveitis may occur concurrently. Fever is common. ESR and CRP are raised. Aspirated joint fluid shows inflammatory changes with a predominance of polymorphonuclear cells. Culture and PCR for *M. bovis*-BCG are negative. Treatment with NSAIDs is usually successful. If there is no response to NSAID treatment, treatment for disseminated BCG infection should be commenced. Intravesical BCG therapy should be permanently discontinued.

Allergic reactions
Allergic reactions with skin rash and arthralgia are rare. If symptoms are mild, treatment with antihistamines and NSAIDs is sufficient, and intravesical therapy

may be continued. If symptoms persist > 48 h, further installations should be postponed until clinical resolution of allergic symptoms.

Severe, persistent allergic symptoms may need to be treated as disseminated BCG reactions, with antimycobacterial drugs, corticosteroids and cessation of intravesical BCG therapy.

8.9.3 Occupational exposure to BCG infection during intravesical therapy

Healthcare workers are at risk for occupational exposure to BCG during preparation and installation of BCG suspension. Splashes to the eye can result in ulceration. Percutaneous exposure may result in localised cutaneous or deep tissue infections with *M. bovis*. A combination of surgical resection and prolonged antimycobacterial therapy may be necessary to cure such infections.

8.10 Antimycobacterial drugs and BCG

The susceptibility of BCG to antimycobacterial drugs varies with the strain. All strains of BCG, like all strains of *M. bovis*, are intrinsically resistant to pyrazinamide.

A study comparing susceptibility of BCG strains to first- and second-line antituberculous drugs found all were susceptible to rifampicin, rifabutin, ethambutol, ciprofloxacin, ofloxacin, streptomycin, amikacin, kanamycin and capreomycin. Strains BCG-Connaught (the strain used in Australia) and BCG-Denmark showed low-level isoniazid resistance and resistance to ethionamide. The clinical significance of low-level isoniazid resistance in BCG strains is unclear. Although this form of resistance in *M. tuberculosis* is not thought to impair the response to standard antituberculous therapy, emergence of rifampicin resistance has been described in a case of disseminated BCG-Denmark infection in an HIV-infected infant treated with a two-drug combination of rifampicin plus isoniazid.

Susceptibility of BCG strains to macrolides is difficult to test, but *in vitro* evidence suggests that strains with a deletion in the RD2 region of the genome (such as BCG-Connaught) are susceptible to macrolides, while those that retain this region (such as BCG-Japan, BCG-Russia, BCG-Sweden and BCG-Moreau) are resistant.

These results suggest that selection of antimycobacterial drugs for BCG infections will be made easier by the clear documentation of the specific BCG strain used for intravesical BCG therapy, and by the isolation, culture and strain-specific identification of acid-fast bacilli from clinical specimens obtained from patients with suspected infectious complications of intravesical BCG therapy.

In keeping with current urological guidelines, we recommend that localised BCG reactions such as granulomatous prostatitis and epididymo-orchitis should be treated with a combination of rifampicin and isoniazid. For disseminated BCG infections, the larger organism burden may lend more weight to the theoretical concern about low-level isoniazid resistance, so addition of ethambutol as recommended by the guidelines (or moxifloxacin if ethambutol is contraindicated) to prevent emergence of resistance is reasonable, albeit unproven.

The additional benefit of adding adjunctive corticosteroids to antimycobacterial drugs for localised and disseminated BCG infection has not been examined in a prospective study, but has a sound theoretical basis and is supported by clinical experience.

References and further reading

Arend SM, van Soolingen D. Low level INH-resistant BCG: A sheep in wolf's clothing? Clin Infect Dis 2011; 52(1):89–93.

Colditz GA, Brewer TF, Berkey CS, et al. Efficacy of BCG vaccine in the prevention of tuberculosis. Meta-analysis of the published literature. JAMA 1994; 271(9):698–702

Gonzalez OY, Musher DM, Brar I, et al. Spectrum of Bacille Calmette-Guerin (BCG) infection after intravesical BCG immunotherapy. Clin Infect Dis 2003; 36(2):140–8.

Goraya JS, Virdi VS. Treatment of Calmette-Guerin bacillus adenitis: A metaanalysis. Ped Infect Dis J 2001; 20:632–4.

Hesseling AC, Schaaf HS, Victor T, et al. Resistant *Mycobacterium bovis* bacillus Calmette-Guerin disease: Implications for management of bacillus Calmette-Guerin disease in human immunodeficiency virus-infected children. Ped Infect Dis J 2004; 23:476–9.

Kolibab K, Derrick S, Morris S. Sensitivity to isoniazid of *Mycobacterium bovis* BCG strains and BCG disseminated disease isolates. J Clin Microbiol 2011; 49:2380–1.

Lamm DL, Blumenstein BA, Crissman JD, et al. Maintenance bacillus Calmette-Guerin immunotherapy for recurrent Ta, T1 and carcinoma in situ transitional cell carcinoma of the bladder: A randomized Southwest Oncology Group study. J Urol 2000; 163(4):1124–9.

Mise K, Goic-Barisic I, Bradaric A, et al. Long-term course and treatment of a cutaneous BCG infection. J Dermatolog Treat 2008; 19(6):333–6.

Ritz N, Tebruegge M, Connell TG, et al. Susceptibility of *Mycobacterium bovis* BCG vaccine strains to antituberculous antibiotics. Antimicrob Agents Chemother 2009; 53(1):316–18.

Witjes JA, Palou J, Soloway M, et al. Clinical practice recommendations for the prevention and management of intravesical therapy-associated adverse events. Eur Urol Supplements 2008; 7(10):667–74.

Index

A

Accelerated BCG response   59, 120
Adherence
   active TB therapy   19
   HIV & TB   87, 89
   intermittent therapy   6
   latent TB therapy   71–72
Adverse drug reactions
   abnormal LFTs   27–30, 72–74
   acute psychosis   32
   arthralgia   31–32
   BCG   *see BCG vaccine*
   corticosteroids   10
   education   19
   gastrointestinal symptoms   27
   hypersensitivity   25–26
   LTBI treatment   72–74
   MAC treatment   101
   peripheral neuropathy   73
   second-line agents   41
   visual disturbance   30–31
Allergy   *see Adverse drug reactions*
Amikacin
   drug-resistant TB   40–47
   MAC   99, 101
   M. abscessus   106
   M. chelonae   105
   M. fortuitum   110
   M. kansasii   104
   M. ulcerans   113
   side effects   41
   use in renal impairment   7
Aminoglycosides   *see also Amikacin*
   use in renal impairment   7

Amoxycillin-clavulanic acid   50
Antiretroviral therapy   *see also HIV and TB*
   drug interactions with TB treatment   85–87
   IRIS   89–90
   regimens with TB treatment   87–89
Antituberculous chemotherapy
   adjunctive steroids   9–13
   adverse effects   *see Adverse drug reactions*
   breast feeding   56
   continuation phase   4, 13, 45–46
   drug-resistant TB   37–49
   initial phase   1–5, 13, 40
   pregnancy   53–54

B

BCG and bladder cancer   122–124
BCG vaccine   115–119
   accelerated response   120
   antimycobacterial drugs   127
   complications   120–121
   contraindications   118
   efficacy   115
   expected response   119
   immune suppression/HIV   118
   indications   116
   local reactions   120
Breast feeding and TB drugs   79
Bronchiectasis   34, 93–96
Bronchoscopy   15, 76, 84, 95–97
Buruli ulcer   *see Mycobacterium ulcerans*

C

Capreomycin 39, 41, 46, 47
Cefoxitin 105–106, 109
Challenge dose 26
Chest X-ray
   in latent TB 75, 82
   in NTM 94–109
   progress after treatment 34
Ciprofloxacin 8, 39, 105, 127
Children & TB 56–58
   BCG 116–120
   MAC 102
Clarithromycin
   MDR-TB 50
   NTM 99–101, 105–107, 109–110, 111, 113
Clofazimine 50
Colour vision testing 19, 20, 31, 100–101
Computerised tomography 84, 95–97
Congenital TB 54, 57–59
Contact tracing 17–19, 74
Continuation phase of TB therapy 4–5, 13–14, 45–46, 87
Corticosteroid use in TB 9–13
Cotrimoxazole
   MDR-TB 50
   NTM 105, 107, 110–111
Cycloserine 8, 9, 40, 45, 48

D

Desensitisation 26
Diabetes mellitus 10, 32, 65, 73
Directly observed therapy 6, 16, 40
Discharge from hospital 21–22
Disseminated disease
   BCG 125–128
   MAC 98
   NTM 105–106
   TB 5, 57, 66, 84

Dosing
   drugs used for MDR-TB 41
   drugs used for NTM 99–100, 103, 106–107, 111, 113
   first-line TB drugs 2
   renal impairment 8
DOTS *see Directly observed therapy*
Doxycycline 107, 110, 105, 111
Drug allergy *see Adverse drug reactions*
Drug interactions 19–20
   antiretroviral therapy 85–86
Drug reactions *see Adverse drug reactions*
Drug-resistant TB 37–50
   assessment 38
   cross/class resistance 42, 47
   MDR-TB *see Multidrug-resistant TB*
   treatment regimens 40–46
   XDR-TB *see Extensively drug-resistant TB*
Drug susceptibility testing
   MAC 99
   MDR-TB 39, 42
   NTM 103–110
Drug monitoring: therapeutic 48
Duration of TB treatment
   drug-resistant 45
   extra-pulmonary, drug-sensitive 5
   HIV 87
   LTBI 77
   pulmonary, drug-sensitive 4

E

Ethambutol
   drug-resistant latent TB 77
   drug-sensitive TB 1–4, 7–9
   hypersensitivity and challenge 25–26
   NTM 98–105, 110–111
   ocular toxicity 30–31

pregnancy 54
resistance 44, 46, 77
use in liver disease 7, 9
use in renal impairment 7, 8
Extensively drug-resistant TB 49–50

F
Fluoroquinolones *see also Moxifloxacin and Ciprofloxacin*
  drug-resistant TB 39, 43–45
  NTM 109–110
  use with renal impairment 8
  XDR-TB 49
Follow up after TB treatment
  LTBI 77–78
  NTM 97, 108, 114
  TB 24–25

G
Genotyping 39
Gout 32

H
Hepatitis *see also Liver disease*
  medication related 27–30, 72–74, 82, 85
  risk groups 23, 27–28, 78, 82, 85
  testing for HBV/HCV 20, 23, 77
HIV and TB 81–89
  antiretroviral regimens 87
  diagnosis of TB 83
  immune reconstitution syndrome 89
  latent TB 82–83
  testing for HIV infection 81
  treatment of TB 84–85, 87
Hospital infection control 17, 18, 21, 58

I
IGRA *see Interferon gamma release assays*

Imipenem 105–106, 109
Immune reconstitution inflammatory syndrome 89–90
Incomplete TB treatment 13
Infertility and TB 62
Initial phase of treatment
  NTM 106, 110
  TB 1–3, 43–47
Inpatient monitoring 20–21
Intensive phase *see Initial phase of treatment*
Interferon gamma release assays 67–71, 82
  HIV 82
  indeterminate results 70
  interpretation 68–69
Intermittent therapy
  MAC 99, 102
  TB 6
Interrupted TB treatment 13
IRIS *see Immune reconstitution inflammatory syndrome*
Isolation 17–18
  drug-resistant TB 40
  mother and neonate 58
Isoniazid
  drug-sensitive TB 1–3, 4–6
  hepatitis 27–30, 72–73, 74
  high dose 43–44, 50
  hypersensitivity and challenge 25–26
  interactions 20
  latent TB 71–74, 82
  M. kansasii 103
  pregnancy 54
  psychosis 32
  resistance 43–44, 45–47
  use in liver disease 7, 9
  use in renal impairment 6, 8

L
Latent TB infection 63–78, 82–83

adherence to therapy 71
HIV 82–83
management 74
pregnancy 61, 77
pre-test probability 64
risk of TB following infection 64
screening indications 74–75
tests and interpretation 65–71
treatment efficacy 71
Linezolid 51
Liver disease *see also Hepatitis*
hepatitis during treatment 27–30, 72–74, 82, 85
pre-existing 7, 9
LTBI *see Latent TB infection*
Lymphadenitis
BCG reaction 120–121
NTM 104
Paradoxical reaction 32

M

MAC infections *see Mycobacterium avium complex*
Mantoux test *see Tuberculin skin test*
MDR-TB *see Multidrug-resistant TB*
Meningitis
corticosteroids 11
duration of therapy 5
in pregnancy 55–57
Microscopy 39, 57, 112
Miliary TB *see Disseminated disease*
Moxifloxacin 3
drug-resistant TB 39–45
drug-resistant latent TB 77
NTM 101, 104, 106, 113
use in renal impairment 8
use in severe liver disease 9
Multidrug-resistant TB 37–50
alternative treatments 50–51
epidemiology 38
laboratory tests 38

post-treatment evaluation 49
risk factors 38
surgery 45
treatment regimens 45–47
Monitoring response to TB therapy 21–22
Mycobacterium abscessus see Rapidly growing mycobacteria
Mycobacterium avium complex 94–102
diagnostic criteria 96
extrapulmonary 102
HIV 95
hypersensitivity-like disease 102
pulmonary infections 94–102
treatment 99–101
Mycobacterium chelonae see Rapidly growing mycobacteria
Mycobacterium fortuitum see Rapidly growing mycobacteria
Mycobacterium kansasii 102
Mycobacterium marinum 110
Mycobacterium ulcerans 111

N

Neonatal TB exposure
management 56–60
Non-tuberculous mycobacteria 93–111
M. avium complex 94–102
M. kansasii 102
M. marinum 110
M. ulcerans 111
rapidly growing mycobacteria 104–110
Notification of TB 18
NTM *see Non-tuberculous mycobacteria*
Nucleic acid amplification tests (NAAT) 39

O

Ocular toxicity 30–31
Outpatient review 23–24

P

Para-aminosalicylic acid   40, 45–48
Paradoxical reaction   32–33, 89
PCR   *see Polymerase chain reaction*
Pericarditis and corticosteroids   10–11
Perinatal management of TB   56–60
   screening for TB   56–60
Peripheral neuropathy   73
Polymerase chain reaction   39
Prednisolone
   indications   9–12
   management of IRIS   90
   rifampicin interaction   19
   risk of reactivation of TB   65
Pregnancy & TB   53–62
   impact of pregnancy on TB   53
   impact of TB on pregnancy   53
   latent TB infection   61
   TB drugs   54
Prothionamide   7, 40–42, 45–48
Pyrazinamide
   arthralgias   31–32
   drug-sensitive TB   1–3, 4–6
   hepatitis   27–30
   hypersensitivity and challenge   25–26
   pregnancy   54–55
   resistance   46–47
   use in liver disease   7, 9
   use in renal impairment   7–8
Pyridoxine   2, 55, 23, 82

Q

Quantiferon-TB Gold In Tube   *see Interferon gamma release assays*
Quinolones   *see Fluoroquinolones*

R

Rapidly growing mycobacteria   104–110
Refugees   10, 64, 74, 76, 79
Renal impairment   30, 41, 65, 73
   drug choice & dose adjustment   6–8
Response to TB treatment   33
RGM   *see Rapidly growing mycobacteria*
Rifabutin
   HIV and active TB   85–86, 88–89
   HIV and latent TB   82
   MAC   99, 101
   MDR-TB   50
Rifampicin
   drug interactions   19
   drug-sensitive TB   1–3, 4–6
   gastrointestinal symptoms   27
   hepatitis   28–30
   hypersensitivity and challenge   25–26
   latent TB   72
   NTM   99–101, 103, 111, 113
   pregnancy   54
   resistance   44, 45–47
   use in liver disease   7, 9
   use in renal impairment   6, 8
Rifapentine   72
Risk of active TB after infection   64

S

Second-line TB medications   39–40
Side effects   *see Adverse drug reactions*
Sputum
   duration of culture positivity in TB   34
   duration of smear positivity in TB   33–34
   monitoring TB treatment   23, 47, 49
   NTM diagnosis   96–97
   PCR   38
   smear positivity and risk of neonatal TB   58, 59
Sterilisation phase   *see Continuation phase*
Steroids   *see Corticosteroid use in TB*
Stevens Johnson syndrome   25

Surgical management
   BCG reactions 121
   lymph node TB 33
   MAC 102
   MDR-TB 50
   NTM 104–112
   pericarditis 10

T

Therapeutic drug monitoring 48–49
Thioridazine 51
TNF α inhibitors *see Tumour necrosis factor α inhibitors*
Transmission, congenital 57–60
Treatment failure 14
Treatment of drug-sensitive TB 1–14
   continuation phase 4
   corticosteroids 9–13
   failure 14
   initial phase 1
   intermittent therapy 6
   interrupted or incomplete treatment 13
   liver disease 7
   renal impairment 6
Tuberculin skin test
   in possible congenital TB 57–61
   interpretation 65–71

Tuberculous meningitis
   corticosteroids for 11
   duration of therapy 5
   in pregnancy 55–57
Tumour necrosis factor α inhibitors
   latent TB reactivation 65, 75–76
   MAC 95

U

Uveitis
   due to BCG 126
   due to rifamycin 99–101

V

Vaccine BCG 115–121
Vestibulo-cochlear toxicity 41, 48
Visual acuity testing 20, 23, 31, 48, 100–101
Visual disturbance 30–31
Vitamin D 20, 77

W

World Health Organization (WHO) 6, 113, 116

X

XDR-TB *see Extensively drug-resistant TB*

www.ingramcontent.com/pod-product-compliance
Lightning Source LLC
Chambersburg PA
CBHW070232180526
45158CB00001BA/409